MEDICAL INTELLIGENCE UNIT

Genetically Modified Mosquitoes for Malaria Control

Christophe Boëte, Ph.D.

Institut de Recherche pour le Développement
Laboratoire Génétique et Evolution des Maladies Infectieuses
UMR CNRS / IRD 2724
Montpellier, France

and

Laboratory of Entomology
Wageningen University and Research Centre
Wageningen, The Netherlands

and

Joint Malaria Programme
Kilimanjaro Christian Medical Centre
Moshi, Kilimanjaro, Tanzania

CRC Press
Taylor & Francis Group
Boca Raton London New York

CRC Press is an imprint of the
Taylor & Francis Group, an **informa** business

GENETICALLY MODIFIED MOSQUITOES FOR MALARIA CONTROL

Medical Intelligence Unit

First published 2006 by Landes Bioscience

Published 2018 by CRC Press
Taylor & Francis Group
6000 Broken Sound Parkway NW, Suite 300
Boca Raton, FL 33487-2742

© 2006 by Taylor & Francis Group, LLC
CRC Press is an imprint of Taylor & Francis Group, an Informa business

First issued in paperback 2019

No claim to original U.S. Government works

ISBN 13: 978-0-367-44630-7 (pbk)
ISBN 13: 978-1-58706-096-0 (hbk)

Visit the Taylor & Francis Web site at
http://www.taylorandfrancis.com

and the CRC Press Web site at
http://www.crcpress.com

While the authors, editors and publisher believe that drug selection and dosage and the specifications and usage of equipment and devices, as set forth in this book, are in accord with current recommendations and practice at the time of publication, they make no warranty, expressed or implied, with respect to material described in this book. In view of the ongoing research, equipment development, changes in governmental regulations and the rapid accumulation of information relating to the biomedical sciences, the reader is urged to carefully review and evaluate the information provided herein.

Library of Congress Cataloging-in-Publication Data

A C.I.P. Catalogue record for this book is available from the Library of Congress.

About the Editor

Christophe Boëte is a Research Scientist at the Institut de Recherche pour le Développement, Montpellier, France. His major research interest lies in the ecology and evolution of host-parasite interactions. In this general area of research his activities fall in several interrelated themes: the evolutionary ecology of the immune response of melanisation of mosquitoes, the malaria parasite's ability to avoid the melanisation response of its invertebrate host, and more recently the impact of vector/parasite interactions on the population biology and community structure of plasmodial species. He is also interested in malaria epidemiology, its control and the public health aspects related to malaria research and the use of traditional medicine against malaria. He received his Ph.D. from Université Pierre et Marie Curie (Paris, France) and did his post-doctoral work as a Marie Curie Fellow at the Wageningen University and Research Centre (Wageningen, The Netherlands) and the Joint Malaria Programme (Moshi, Tanzania).

CONTENTS

RESEARCH ORIENTATIONS AND ETHICAL ISSUES CONCERNING THE USE OF GM MOSQUITOES FOR MALARIA CONTROL

CONCLUSION

EDITOR

Christophe Boëte
Institut de Recherche pour le Développement
Laboratoire Génétique et Evolution des Maladies Infectieuses
UMR CNRS / IRD 2724
Montpellier, France
and
Laboratory of Entomology
Wageningen University and Research Centre
Wageningen, The Netherlands
and
Joint Malaria Programme
Kilimanjaro Christian Medical Centre
Moshi, Kilimanjaro, Tanzania
Email: cboete@gmail.com
Introduction, Chapters 7, 14

CONTRIBUTORS

Anthony E. Brown
Genetics Unit
Department of Biochemistry
University of Oxford
Oxford, U.K.
Chapter 3

Diarmid H. Campbell-Lendrum
Disease Control and Vector Biology Unit
Department of Infectious
    and Tropical Diseases
London, U.K.
and
Occupational and Human Health Team
World Health Organization
Geneva, Switzerland
Chapter 6

Julie Castro
Comité pour l'Annulation de la Dette
    du Tiers-Monde France (CADTM)
Jargeau, France
Chapter 2

Flaminia Catteruccia
Department of Biological Sciences
Imperial College London
London, U.K.
Chapter 3

Christine Chevillon
Team 'Evolution des Systèmes
    Symbiotiques'
Laboratoire Génétique et Evolution
    des Maladies Infectieuses
Institut de Recherche pour le
    Développement
Montpellier, France
Chapter 10

Paul G. Coleman
Disease Control and Vector Biology Unit
Department of Infectious
    and Tropical Diseases
London, U.K.
and
Oxitec Limited
Abingdon, Oxford, U.K.
Chapter 6

Carlo Costantini
Institut de Recherche
  pour le Développement
Centre IRD de Ouagadougou
Ouagadougou, Burkina Faso
Chapter 11

Andrea Crisanti
Department of Biological Sciences
Imperial College London
London, U.K.
Chapter 3

Christopher Curtis
London School of Hygiene
  and Tropical Medicine
London, U.K.
Chapter 6

Thierry de Meeûs
Team 'Evolution des Systèmes
  Symbiotiques'
Laboratoire Génétique et Evolution
  des Maladies Infectieuses
Institut de Recherche
  pour le Développement
Montpellier, France
Chapter 10

Mahamadou Diakité
Malaria Research and Training Center
Department of Epidemiology
  of Parasitic Diseases
Faculty of Medicine, Pharmacy
  and Odonto-Stomatology
University of Bamako
Bamako, Mali
Chapter 1

Alassane Dicko
Malaria Research and Training Center
Department of Epidemiology
  of Parasitic Diseases
Faculty of Medicine, Pharmacy
  and Odonto-Stomatology
University of Bamako
Bamako, Mali
Chapter 1

Abdoulaye A. Djimdé
Malaria Research and Training Center
Department of Epidemiology
  of Parasitic Diseases
Faculty of Medicine, Pharmacy
  and Odonto-Stomatology
University of Bamako
Bamako, Mali
Chapter 1

Amagana Dolo
Malaria Research and Training Center
Department of Epidemiology
  of Parasitic Diseases
Faculty of Medicine, Pharmacy
  and Odonto-Stomatology
University of Bamako
Bamako, Mali
Chapter 1

Ogobara K. Doumbo
Malaria Research and Training Center
Department of Epidemiology
  of Parasitic Diseases
Faculty of Medicine, Pharmacy
  and Odonto-Stomatology
University of Bamako
Bamako, Mali
Chapter 1

Chris Drakeley
Department of Infectious
  and Tropical Diseases
London School of Hygiene
  and Tropical Medicine
London, U.K.
and
Joint Malaria Programme
Kilimanjaro Christian Medical Centre
Moshi, Kilimanjaro, Tanzania
Chapter 8

Heather Ferguson
Public Health Entomology Unit
Ifakara Health Research
  and Development Centre
Ifakara, Tanzania
and
Laboratory of Entomology
Wageningen University
Wageningen, The Netherlands
Chapter 9

Sylvain Gandon
Laboratoire Génétique et Evolution
  des Maladies Infectieuses
Institut de Recherche
  pour le Développement
Montpellier, France
Chapter 9

Kassoum Kayentao
Malaria Research and Training Center
Department of Epidemiology
  of Parasitic Diseases
Faculty of Medicine, Pharmacy
  and Odonto-Stomatology
University of Bamako
Bamako, Mali
Chapter 1

David W. Kelly
Disease Control and Vector Biology Unit
Department of Infectious
  and Tropical Diseases
London, U.K.
and
Oxitec Limited
Abingdon, Oxford, U.K.
and
Department of Zoology
University of Oxford
Oxford, U.K.
Chapter 6

Bart G.J. Knols
International Atomic Energy
  Agency (IAEA)
Agency's Laboratories in Seibersdorf
Seibersdorf, Austria
Chapter 12

Tovi Lehmann
Laboratory of Malaria
  and Vector Research
National Institute of Allergy
  and Infectious Diseases (NIAID)
National Institutes of Health (NIH)
Rockville, Maryland, U.S.A.
Chapter 5

Tom J. Little
Institute of Evolutionary Biology
School of Biology
University of Edinburgh
Edinburgh, Scotland, U.K.
Chapter 4

Darryl Macer
RUSHSAP, UNESCO Bangkok
Prakanong, Bangkok, Thailand
Chapter 13

Margaret Mackinnon
Department of Pathology
University of Cambridge
Cambridge, U.K.
Chapter 9

Damien Millet
Comité pour l'Annulation de la Dette
  du Tiers-Monde France (CADTM)
Jargeau, France
Chapter 2

Richard E. Paul
Laboratoire d'Entomologie Médicale
Institut Pasteur de Dakar
Dakar, Sénégal
Chapter 10

Elisa Petris
Department of Biological Sciences
Imperial College London
London, U.K.
Chapter 3

Andrew Read
Institute of Immunity
  and Infection Research
University of Edinburgh
Edinburgh, Scotland, U.K.
Chapter 9

François Renaud
Team 'Evolution des Systèmes
  Symbiotiques'
Laboratoire Génétique et Evolution
  des Maladies Infectieuses
Institut de Recherche
  pour le Développement
Montpellier, France
Chapter 10

Hugh Reyburn
Department of Infectious
  and Tropical Diseases
London School of Hygiene
  and Tropical Medicine
London, U.K.
and
Joint Malaria Programme
Kilimanjaro Christian Medical Centre
Moshi, Kilimanjaro, Tanzania
Chapter 8

Christina Scali
Department of Biological Sciences
Imperial College London
London, U.K.
Chapter 3

Frédéric Simard
Institut de Recherche
  pour le Développement
Unité de Recherche 016
Laboratoire de Recherche
  sur le Paludisme
Organisation de Coordination
  pour la lutte Contre les Endémies
  en Afrique Centrale (OCEAC)
Yaoundé, Cameroun
Chapter 5

Willem Takken
Laboratory of Entomology
Wageningen University
Wageningen, The Netherlands
Chapter 11

Mahamadou A. Théra
Malaria Research and Training Center
Department of Epidemiology
  of Parasitic Diseases
Faculty of Medicine, Pharmacy
  and Odonto-Stomatology
University of Bamako
Bamako, Mali
Chapter 1

Yeya T. Touré
Special Programme for Research and
  Training in Tropical Diseases (TDR)
World Health Organization
Geneva, Switzerland
Chapter 12

Boubacar Traoré
Malaria Research and Training Center
Department of Epidemiology
  of Parasitic Diseases
Faculty of Medicine, Pharmacy
  and Odonto-Stomatology
University of Bamako
Bamako, Mali
Chapter 1

══ INTRODUCTION ══

Everything (or Almost Everything) You Want to Know about Genetically Modified Mosquitoes for Malaria Control but Are (Maybe) Afraid to Ask

Despite a century of research and attempts to control one of the deadliest foes of mankind, the malaria situation remains a major public health problem.[1] Obviously biological explanations (the resistance of parasites and mosquitoes against available drugs and insecticides respectively) are often given, but they remain partial and incomplete. Indeed, the deterioration of socio-economic conditions due to the policies imposed on many developing countries by international financial institutions, such as the structural adjustment programmes and the mechanism of debt, plays an important role in the malaria situation and its evolution.[2]

In the last decade, molecular biology has been a source of great hope for creating genetically-modified mosquitoes able to resist the malaria parasite.[3] If technical progress permits confidence in the creation of such non-vectors, many questions remain open concerning the putative success of their deployment and the resultant reduction of malaria transmission. Indeed the understanding of the coevolutionary processes underlying malaria/mosquito interactions is crucially lacking despite its enormous importance.[4] Moreover, when discussing transgenic mosquitoes, one critical point is the spread of the allele conferring resistance in mosquito populations ensuring the replacement of one or several populations of vectors able to transmit malaria by (theoretically) unable one(s). However, invading a whole population of mosquitoes with a transgene (composed with an allele conferring malaria-resistance and a driving system) is unlikely to be an easy task, it will at least depend on the population structure[5] and on the quality of the driver.[6] Alongside this, it appears that the spread of refractoriness itself is necessary but not sufficient as interactions between the allele of interest, the parasite and the environment may affect refractoriness[7] and thus limit the expected success in terms of malaria control. Indeed the aim of a release of transgenic mosquitoes is not the spread of an allele of interest in mosquito populations but a real decrease in the malaria burden, it seems then crucial to have a look at the possible consequences of such a release. How does a reduction in malaria transmission affect the epidemiology of the disease?[8] What could be the evolutionary consequences in terms of the virulence of the parasite?[9] Thus it appears that the idea of using GM mosquitoes opens up more questions than answers and calls for some rethinking in malaria biology.[10] Finally, if the mainstream perspective concerning the use of transgenic mosquitoes is dealing with spreading refractoriness in wild populations of mosquitoes, little has been done about methods for affecting the mosquito-host interactions whether it be with GM technology or using more conventional methods.[11] However, prior to any release of transgenic insects, numerous ethical, legal and social questions[12,13] are still pending and questioning the interest of such a high-tech method for malaria control and its societal implications seems highly necessary.[14]

Christophe Boëte, Ph.D.

References

1. Théra MA, Djimde AA, Dicko A et al. Malaria situation in the beginning of the 21st century. In: Boëte C, ed. Genetically Modified Mosquitoes for Malaria Control. Georgetown: Landes Bioscience, 2006; 1:1-15.
2. Castro J, Millet D. Malaria and structural adjustment: Proof by contradiction. In: Boëte C, ed. Genetically Modified Mosquitoes for Malaria Control. Georgetown: Landes Bioscience, 2006; 2:16-23.
3. Catteruccia F, Brown AE, Petris E et al. Development of a toolkit for manipulating malaria vectors. In: Boëte C, ed. Genetically Modified Mosquitoes for Malaria Control. Georgetown: Landes Bioscience, 2006; 3:24-35.
4. Little TJ. Immune system polymorphism: Implications for genetic engineering. In: Boëte C, ed. Genetically Modified Mosquitoes for Malaria Control. Georgetown: Landes Bioscience, 2006; 4:36-45.
5. Simard F, Lehmann T. Predicting the spread of transgene in African malaria vector populations: Current knowledge and limitations. In: Boëte C, ed. Genetically Modified Mosquitoes for Malaria Control. Georgetown: Landes Bioscience, 2006; 5:46-59.
6. Curtis CF, Coleman PG, Kelly DW et al. Advantages and limitations of transgenic vector control: Sterile males versus genes drivers. In: Boëte C, ed. Genetically Modified Mosquitoes for Malaria Control. Georgetown: Landes Bioscience, 2006; 6:60-78.
7. Boëte C. Malaria-refractoriness in mosquito: Just a matter of harbouring genes? In: Boëte C, ed. Genetically Modified Mosquitoes for Malaria Control. Georgetown: Landes Bioscience, 2006; 7:79-88.
8. Reyburn H, Drakeley C. The epidemiological consequences of reducing the transmission intensity of P. falciparum. In: Boëte C, ed. Genetically Modified Mosquitoes for Malaria Control. Georgetown: Landes Bioscience, 2006; 8:89-102.

9. Ferguson HM, Gandon S, Mackinnon MJ et al. Malaria parasite virulence in mosquitoes and its implications for the introduction and efficacy of GMM malaria control programmes. In: Boëte C, ed. Genetically Modified Mosquitoes for Malaria Control. Georgetown: Landes Bioscience, 2006; 9:103-116.
10. Chevillon C, Paul RE, de Meeûs T et al. Thinking transgenic vectors in a population context: Some expectations and many open-questions. In: Boëte C, ed. Genetically Modified Mosquitoes for Malaria Control. Georgetown: Landes Bioscience, 2006; 10:117-136.
11. Takken W, Costantini C. The genetic of vector-host interactions: Alternative strategies for genetic engineering for malaria control. In: Boëte C, ed. Genetically Modified Mosquitoes for Malaria Control. Georgetown: Landes Bioscience, 2006; 11:137-145.
12. Touré YT, Knols BGJ. Genetically-modified mosquitoes for malaria control: Requirements to be considered before field releases. In: Boëte C, ed. Genetically Modified Mosquitoes for Malaria Control. Georgetown: Landes Bioscience, 2006; 12:146-151.
13. Macer D. Ethics and community engagement for GM insect vector release. In: Boëte C, ed. Genetically Modified Mosquitoes for Malaria Control. Georgetown: Landes Bioscience, 2006; 13:152-165.
14. Boëte C. Transgenic mosquitoes for malaria control: Time to spread out of the scientific arena. In: Boëte C, ed. Genetically Modified Mosquitoes for Malaria Control. Georgetown: Landes Bioscience, 2006; 14:166-170.

CHAPTER 1

Malaria Situation in the Beginning of the 21st Century

Mahamadou A. Théra, Abdoulaye A. Djimdé, Alassane Dicko,
Mahamadou Diakité, Kassoum Kayentao, Boubacar Traoré,
Amagana Dolo and Ogobara K. Doumbo*

Introduction

Falciparum malaria parasite, an avian-originated parasite, has probably coevolved with human being (*Homo sapiens*) since the discovery of agriculture, around 20-30 000 years ago. The very devastating parasite disease has spread worldwide and killed millions of people. This ancient disease became the subject of intensive research efforts when malaria stood as an important obstacle to the expansion of the colonial army in Africa and malaria endemic area in South East Asia and America. As a result, major progress in the understanding and control of malaria was achieved during the 19th and the 20th centuries. The causative agent of malaria was discovered in 1880 by Laveran.[1] A dramatic advance in dissecting the life cycle of malaria was achieved when Ronald Ross (working in India), Mico Bignami and Giuseppe Bastianeli (working in Italy) showed in the late 1890s that mosquitoes transmitted the parasite.[2,3] The discoveries of a very potent insecticide dichloro-diphenyl trichloroethane (DDT) and an extremely efficacious drug, chloroquine, generated much hope in the possible eradication of malaria and prompted the launch of an ambitious program for the worldwide eradication of malaria by WHO in 1955.[4] The Program to eradicate malaria using DDT and chloroquine met with successes in some countries (mostly industrialized countries and in areas where ecological conditions were less favorable to the anopheline vectors).[5] However, largely because of the development of mosquito resistance to DDT and the appearance and spread of *P. falciparum* chloroquine resistance, the eradication program was abandoned in 1969.[6] Furthermore, Sub-Saharan Africa countries with *Anopheles gambiae s.l.*, the most competent vector and an environment particularly favorable to malaria transmission, were excluded from the eradication campaign.[5] Subsequently, Sub-Saharan Africa has sustained the major burden of malaria morbidity. Today, malariologists and health officials more modestly talk about "controlling" the disease. WHO encourages and supports the creation of Malaria Control Programs in endemic countries. The main goal is now to decrease the mortality and morbidity due to malaria. The available tools include insecticide impregnated bed nets and a shrinking number of effective antimalarial drugs.

When malaria affected industrialized countries including the United States and Europe, it was among the best-studied infections.[7] As the disease became eradicated or mostly controlled

*Corresponding Author: Ogobara K. Doumbo—Malaria Research and Training Center,
Department of Epidemiology of Parasitic Diseases, Faculty of Medicine, Pharmacy
and Odonto-Stomatology, University of Bamako, BP 1805, Bamako, Mali.
Email: okd@mrtcbko.org or okd@ikatelnet.net

Genetically Modified Mosquitoes for Malaria Control, edited by Christophe Boëte.
©2006 Landes Bioscience.

in the most affluent countries, support for malaria research also diminished considerably. Because of market forces, very few major pharmaceutical companies have active programs for the development of new antimalarial compounds.[8]

Today malaria is reemerging in some of the places where it had been eradicated. And because ease of travel and the global economy are bringing the malarious areas closer than ever to the rest of the world, there is an increasing mobilization of resources for malaria research and control. Several new programs such as Roll Back Malaria (RBM), the Multilateral Initiative on Malaria (MIM), the WHO/World Bank/UNDP Special Program on Tropical Disease Research (TDR), the European Malaria Vaccine Initiative (EMVI), the Malaria Vaccine Initiative of the Bill and Melinda Gates Foundation, Medicines for Malaria Venture (MMV), the European and Developing Countries Clinical Trials Platform (EDCTP), have been launched in the past few years and are drawing much needed resources into malaria research. Although an actual impact of these programs on the daily lives of people suffering from malaria is still not satisfactory, they carry much hope for a better future.

At the outset of the 21st century, malaria and other reemerging diseases are still posing the greatest threat to human health, leading to the fatalistic feeling that there has been no change for 200 years.

In this review we will explore the indicators used to assess the burden of malaria, review the actual numbers on disease burden, and the specific target population at most risk for malaria. We will then discuss the actual major control strategies and the challenge raised by parasite resistance to anti malarial drugs. Further on, we will discuss the role of genetic factors on malarial infection and malaria-induced immune responses, including different interaction with different ethnic groups. And finally we will review perspectives for malaria control, focusing on new areas of research offered by genomics and discuss the potential, hurdles and hopes for a malaria vaccine. As African and researching new tools for malaria control, we believe that the eradication campaign was a success where it was applied. A more powerful international alliance in research for new tools and control for efficient application of existing tools will be needed to move the parasite from its preferred environment in Sub-Saharan Africa. The only way to significantly reduce malaria-related morbidity, mortality, neurological sequelae and socio-economic burden in endemic countries is to achieve at least 80% coverage rate with the existing control measures in their population. This will require a strong international commitment and trustful partnerships at all levels.

Indicators and Malaria Burden

Malaria Indicators Outside Pregnancy

Clear, simple and meaningful indicators are important to describe and characterize malaria morbidity and mortality and to assess the impact of control measures. Three types of epidemiological settings are used to broadly describe the global situation of malaria: area with stable endemic malaria, area with unstable malaria and area free of malaria.

Several indicators have been used to describe malaria burden in endemic areas. These indicators are expressed in terms of prevalence or incidence, using clinical, parasitological and molecular data or their combinations.

The prevalence of enlarged spleen (also called spleen index) in children of 2-9 years of age proposed by the World Health Organization[9] were first been used to classify the level of malaria endemicity.[5] Similar classification was proposed later using the prevalence of malaria parasite in blood smears from children aged 2-9 years.[10,11]

A major advantage of these two indicators is the fact that they are easy to measure and standardize. The fact that they do not always correlate with malaria morbidity and mortality and the dynamic of malaria transmission in the overall population, constitute their major disadvantage. Overall malaria morbidity may be higher in a lower endemic area than in areas where malaria is hyperendemic due to the late onset of acquired immunity.[12-15] Therefore neither of these indicators was recommended for the evaluation of malaria control efforts.

Successful interventions might well reduce the morbidity and mortality, without immediate direct impact on parasite index or spleen index. Age specific incidence of malaria disease and death from malaria provide a better estimation of malaria burden in an area and better reflect the impacts of control efforts.

Malaria disease can be divided in two forms: uncomplicated malaria and severe malaria. Severe malaria as defined by the WHO refers to person with asexual form of malaria parasite with one of the following criteria: prostration, impaired consciousness, respiratory distress or pulmonary edema, seizure, circulatory collapse, abnormal bleeding, jaundice, hypoglycemia, hyperparasitemia, hemoglobinuria or severe anemia (defined as hemoglobin <5 g/dL or hematocrit <15%).[16] Majority of severe malaria cases are cerebral malaria and severe anemia or a combination of the two.

Uncomplicated malaria is usually defined as the presence of parasites in the blood smear associated with symptoms and signs consistent with malaria with none of the signs or symptoms of severity listed above. The entity is difficult to define in malaria endemic area because parasitemia is very common and the signs and symptoms are not specific to malaria. Most common symptoms include fever, chills, sweating, headaches, muscle pains, spleen enlargement nausea and vomiting.[17-19] Because fever is the most common sign and temperature can be measured and quantified easily, fever alone and fever plus parasitemia have been used to define presumptive malaria and clinical malaria respectively. Statistical models to determine threshold parasite density for malaria disease definition, which also allows computing the attributable fraction of malaria, were proposed.[20,21] While fever increases with increased malaria parasite density in blood smears, different threshold of parasite density for malaria fever have been reported depending on age, and the transmission intensity.[22-25]

The incidence of severe malaria in the other hand is much easier to define. The WHO definition is the most widely used. The clinical presentations of severe malaria morbidity vary with age and transmission intensity. Severe anemia is more common in younger children with age of three years and less as compared to those in older age groups and in area of high perennial transmission.[26,27] Most commonly used indicators of malaria mortality include: malaria specific mortality, malaria proportional mortality and malaria case fatality.

In addition to the above indicators of malaria burden, recently environmental indicators have been proposed to describe the pattern of malaria transmission. The new tools of remote sensing (RS) and Geographical Information System (GIS) served to integrate information on various environmental factors including climate and altitude. This approach was used successfully by the MARA (Mapping Malaria Risk in Africa, www.mara.org.za/trview_f.htm, accessed on October 10, 2005) project to map the different levels of malaria transmission and also served to predict rather reliably, occurrence of malaria epidemics in given areas.

Malaria Indicators during Pregnancy

If the above-described indicators are useful to describe the burden of malaria in children, they fail to fully capture the burden of malaria during pregnancy. Malaria affects pregnant women depending on the intensity of transmission, the mother's immunity to malaria infection and the number of previous pregnancies. In areas of low transmission, pregnant women are more susceptible to acute malaria, which may result in spontaneous abortions, stills birth, death of the mother. Chronic manifestations such as anemia and placental malaria are less common.[28] In areas of stable transmission, pregnant women are more likely to develop chronic disease such as anemia, placental infection which both may subsequently lead to low birth weight,[29,30] which in turn is an important contributor to neonatal mortality.[31] Indicators of malaria during pregnancy should include anemia. Mild anemia causes prematurity and fetal intra-uterine growth retardation, while severe anemia may lead to maternal and fetal mortality, women being more at risk during their first pregnancy. Placenta is a preferential setting for parasites sequestration and development.[32] Parasite multiplication within the syncitiotrophoblast alters placenta structure and leads to decreased flow of oxygen and nutrient from the mother. This may provoke abortion, fetus grow retardation, low birth weight and prematurity which are important risk factors for infant mortality.[33,34] Therefore, in addition to anemia, placental infection rather than presence of parasites in peripheral blood and LBW should be used as indicators to assess the impact of control measures targeting malaria during pregnancy.

Disease Burden

Because of poor health systems in Sub-Saharan Africa, the real burden of malaria is not known.[35] Using the best available source of data and novel analysis methods,[36-38] recent updated numbers on the burden of malaria acknowledged previous gross underestimation of the toll Africans are paying to the plague. In 2000, it was estimated that 115,750,109 [95% CI: 91,242,971-257,956,670] episodes of malaria fever occurred among the 96 millions of under-five children exposed in Sub-Saharan Africa.[38] Among those, 544,427 cases [95% CI: 95,513-1,757,448] evolved to become severe malaria episodes and were admitted to hospital for treatment. Given the very low attendance rate of clinics, the actual numbers of severe malaria could be as high as double of the upper limit of the uncertainty interval. Malaria specific mortality was estimated to be 803,000 [95% CI: 709,855-896,145] in 2000, in Sub-Saharan Africa.[39] Direct and indirect cost of malaria in Africa was estimated at more than $2,000 millions, malaria being a major cause of economic and social poverty.[35,40] Each year, 23 million pregnancies are at risk of malaria in Sub-Saharan Africa.[41]

The strengthening of health care systems and systematic routine documentation of morbidity and mortality will allow more precise estimation of malaria burden in Sub-Saharan Africa.

Malaria Control

The objectives of current malaria control strategies are to reduce mortality and morbidity due to malaria. Appropriate strategies vary according to the epidemiological setting and the resources available. The global strategy for malaria control adopted at the WHO ministerial conference in October 1992[42] recommended: (1) Early diagnostic and prompt treatment, (2) Planning and implementation of preventive, selective and sustainable vector control measures such as use of insecticide impregnated materials, (3) Detection and prevention of epidemics, (4) Strengthening local capacities in basic and applied research to permit and promote the regular assessment of the country's malaria situation, in particular ecological, social, and economical determinants of the disease. Because the malaria situation varies from country to country and even within the same country controls strategies should custom-made based on local epidemiological characteristics.

Our review will mainly focus on issues related to antimalarial treatment and innovative preventive approaches for pregnant women. Issues related to vector control measures are addressed elsewhere in this book.

Treatment is a critical component of malaria control. It prevents death from malaria disease. The efficacy of this strategy depends largely on the drug efficacy and availability. The resistance of *Plasmodium falciparum* to chloroquine and other affordable drugs such as sulfadoxine-pyrimethamine and amodiaquine constitute one of the most serious threats to malaria control. Trape et al[43] have reported an increase in malaria mortality, thought to be primary due to the parasite resistance to chloroquine. Resistance to these affordable drugs has triggered important efforts to develop new drugs and to combine antimalarials so that they are more effective and delay the development of resistance. In addition tremendous efforts have explored the mechanisms of resistance to drugs with the aims to develop tools to easily monitor the dynamics of the resistance phenomenon worldwide. Molecular markers for monitoring of resistance to chloroquine and to sulfadoxine-pyrimethamine, the most affordable antimalarial drugs, have been proposed.[44-49] Such markers can be extremely useful tools for national malaria control program and allow evidence-based decision making with regards to antimalarial drug policy.

WHO now recommends combination therapy as treatment policy for falciparum malaria in all countries with resistance to choloroquine and/or sulfadoxine-pyrimethamine. Use of combinations of antimalarials to treat malaria disease not only is effective but also delays the development of the resistance of malaria parasites to these drugs. Artemisinin-based combination therapies (ACTs) are highly efficacious drugs against choloroquine resistant falciparum malaria in various regions[50-52] offer an additional benefit of reducing the gametocytes carriage and possibly malaria transmission.[53,54] According to the World Malaria Report (2005) more than 40 malaria-endemic countries (including 23 African countries) have adopted ACTs. It is expected that more countries will adopt the ACTs in the next few years. However, even though, ACTs are very efficacious and are being adopted as first or second line antimalarial therapies in endemic countries with resistant falciparum malaria, many challenges remain to insure prompt and effective treatment of malaria in malaria endemic areas. Current challenges include: (1) the production of sufficient quantity of ACTs under current Good Manufacturing Practices, to treat all clinical malaria cases, (2) the delivery and availability of these drugs where needed (including in rural areas) at a cost that can be supported by populations in malaria endemic areas. Potential problems include the possibility of unknown side effect with large and repeated usage of the ACTs particularly in African countries and inappropriate use particularly during the first trimester of the pregnancy.

Intermittent Preventive Treatment of Malaria (IPT)

Several studies have demonstrated the effectiveness of the administration of a curative dose of antimalarial at set time intervals in preventing malaria during pregnancy.[29,55,56] However, a Tanzanian study showed that pregnant women were concerned about using sulfadoxine-pyrimethamine as IPT during their pregnancy and this highlights the need of careful assessment of local situations before large implementation of IPT.[57] World Health Organization currently recommends the use of IPT for pregnant women in all areas with stable transmission of falciparum. It is now introduced in 26 countries in Africa (World Malaria Report 2005). Recently, controlled clinical trials conducted in infants and children in area of high perennial or seasonal transmission showed that intermittent preventive treatment reduced the incidence of first malaria episode and severe anemia by more than 50%,[58,59] was simple and should be relatively easy to implement. Additional research on the best antimalarial to use, the best treatment schedule, possible rebound effect and impact on the development of resistance of the malaria parasites is needed.[60]

Genetic Susceptibility and Malaria

Experimental models in animals are of great value for the initial identification and functional analysis of complex disease genes. But the final evidence for the involvement of these genes in human diseases such as malaria must come from extensive genomic epidemiological studies carried-out in several populations at most risk of the disease. African populations are the most genetically diverse in the world.[61] This means that the study of the genetic basis of resistance and susceptibility to important diseases, such as malaria, may be particularly rewarding in Africa. In addition, Africa experiences many ethnic groups, between these, some (e.g., Fula [or Peulh] in Burkina Faso and in Mali) are more resistant to malaria compared to other ethnic groups.[62,63] A key objective of research in human genomics is not only for medical purposes, but also, for the basic understanding of malaria parasites and human coevolution.

One of the most complex issues of human genomics in malaria is to quantify the contribution of host genetic determinants in the variation in susceptibility to disease. Estimates of the genetic contribution, that are relatively independent of environmental factors, have been obtained from studies on twins. The risk of developing certain diseases (malaria, tuberculosis, HIV/AIDS) showed a significant heritable component by comparing homozygous twins (who are genetically identical) with dizygous twins (who are genetically related but not identical). Such observations, in addition to several molecular data, have led to the view that genetic factors play a role in almost all-human disease, even if the primary cause is environmental.[64,65] Over the past few years, notable progress has been made using genetics as a tool towards the discovery of loci encoding host susceptibility/resistance to malaria.[66,67] Natural selection serves to filter best fitted inherited genetic variation between individuals resulting in different ability to survive and reproduce successfully. In malaria endemic regions, most children present the mild form of the disease. Only a small percentage of those infected develop severe or complicated disease and consequently die of it. This is mainly explained by host resistance factors that have evolved over several thousand years of selection under the pressure of high exposure to falciparum malaria.[68] Thus, genes that confer resistance to severe malaria are widespread in populations in endemic areas such as Sub-Saharan Africa.

Red Blood Cells Polymorphism and Malaria

The coevolution of the parasite within its human ecological niche is the major factor that generated polymorphisms in human red blood cell (RBC). Red blood cells polymorphisms associated with malaria are structural changes of the β chain (β) of hemoglobin (type S, C, E or F); abnormalities of globins chain synthesis (α and β thalassemia); deficiencies in red blood cells enzymes (glucose-6-phosphate-dehydrogenase) and changes at the membrane of red blood cell (Duffy antigens and ovalocutosis).[69] WHO estimated that 5% of world population carries abnormalities of red blood cells.[70] More than 50 years ago, Haldane suggested that protection against malaria was conferred to individuals by a form of thalassemia. Since, the observed overlap in the geographic distribution of malaria with hemoglobinopathies and other red blood cell disorders have been cited in support of the hypothesis that malaria has been an important evolutionary force in the selection of these variants. Epidemiology and in vitro support for the malaria hypothesis is best documented for the thalassemia[71] and sickle hemoglobin (HbS) within different regions of Africa, the Middle East, and Asia.[66,72] Recent studies showed that HbC is associated with protection from mild malaria in Burkina Faso[73,74] and from severe malaria in the Dogon of Mali in West Africa.[75] Fairhurst et al,[76] showed that Hemoglobin C might protect against malaria by reducing PfEMP-1 (*P. falciparum* erythrocyte membrane protein-1) mediated adherence of parasitized erythrocytes, thereby mitigating the effects of their sequestration in the microvasculature. Ruwende et al[77] reported that, in two large case-control studies of over 2,000 African children, the common African form of G6PD deficiency (G6PD A-) is associated with a 46-58% reduction in risk of severe malaria. The mechanisms of protection against malaria are not well known.

HLA Type and Malaria

Piazza et al[78] were the first to present evidence of the association between particular HLA variants and malaria in Sardinia Island (Italy). They compared lowland areas where malaria occurred and highland areas. Since then a case-control study in the Gambia indicated that the HLA class I antigen HLA-Bw53 and the HLA class II haplotypes DRB1*1302-DRB1*0501 both protect against severe malaria.[66] Furthermore, in population studies, these genotypes account for as great a reduction in disease incidence as the sickle cell polymorphism. Therefore, they confer 40% reduction in life-threatening complications of malaria in Gambian children. Many studies have shown the role of other host genetic factors in susceptibility/resistance to malaria infection. There is much interest in a group of SNPs located at nucleotides -238, -308, -376 with respect to the TNF transcriptional start site; all are substitution of adenine for guanine.[79] The association of -308A homozygote with cerebral malaria was found in Gambian children,[79] and a similar tendency was observed in Kenyan children.[80] Since, some contradicting results have been found with these SNPs in different studies on severe malaria.

In addition, CD36 is a host receptor, involved in *P. falciparum* cytoadherence, expressed on the endothelium, platelets and leucocytes. Recently its involvement in the immune response to malaria has been proposed by two distinct studies. Binding of infected erythrocytes to dendritic cells, via CD36, inhibited the maturation of these cells and reduced their capacity to stimulate T cell.[81] On the other hand, a role for CD36 in parasite clearance was proposed.[82] African populations contain a high frequency of nonsense and frame-shift mutations in the gene coding for CD36 and divergent results have been produced concerning the effect of these mutations on malaria susceptibility.[83] Aitman et al[84] observed that an exceptionally high frequency of the heterozygous or the homozygous state for different CD36 truncation mutations is significantly associated with susceptibility to malaria, particularly cerebral malaria. While, Pain et al[85] have shown that a nonsense mutation (exon 10 188 T→G) is associated with protection from severe disease but not from cerebral malaria. Similarly, polymorphisms in CD36 promoter regions (CD36-14T→C and CD36-53G→T) were significantly decreased in cerebral malaria compared to mild malaria.[86] These discrepancies probably reflect the contrasting roles that this molecule seems to have in malaria pathogenesis. As suggested by Pain,[85] the molecular dissection of the effects that gene variability exert on the biological reactions mediated by CD36 could give some clues on the implications of the associations found.

Unfortunately, the discovery of these polymorphisms has not led to the development of new treatments or prophylaxis against malaria; most of these mutations carry serious consequences for the host when homozygous and therefore have been offered no practical therapeutic lead. The hypothesis underlying malaria is that its susceptibility may be determined by many hundreds of different genetic polymorphisms and that disease severity lies not in the individual polymorphism, but in what it tells us about the control of immune processes in general.

Despite some progress in this area, it will still be important to view the study of human genetic disease from an epidemiological perspective. As opposed to experimental science, both human genetics and epidemiology are observational sciences. We will never be able to exert the same degree of scientific control in studying human disease that experimentalist can obtain with model systems. Furthermore, we must not lose sight of the numerous nongenetic factors that influence malaria risk, and how they interact with host factors. The genetic approach aims to correlate differences in disease frequencies between groups with different polymorphism frequencies.

Humoral Immune Factors

The interethnic differences with regard to the humoral immune responses against a crude *P. falciparum* blood stage antigen were evaluated in individuals of Dogon and Fulani's ethnic groups in Mali. The sera levels of anti-*P. falciparum* IgE and anti-*P. falciparum* IgG antibodies in the Fulani were significantly higher than in the Dogon.[62] For total IgE there was a trend of higher levels in the Fulani. However, this did not reach statistical significance. Then, natural antibodies to malaria specific antigens were tested. The Apical Membrane Antigen (AMA-1) and the Merozoite Surface Antigen (MSP-1) of *P. falciparum*, two candidate malaria vaccines antigens were chosen. Geometric means of anti AMA-1(FVO and3D7) and anti MSP-1 (FVO and 3D7) antibody titers were significantly higher in Fulani than in Dogon. Bolad et al, 2005[87] compared levels of the IgG subclasses of the *P. falciparum* reactive antibodies in the sympatric tribes in Burkina Faso (Fulani/Mossi) and Mali (Fulani/Dogon). Anti-malarial IgG and IgM as well as IgG1 and IgG3 antibodies were consistently significantly higher in the Fulani as compared to the nonFulani, in both countries. Taken together the elevated levels of IgG1 and IgG3 antibodies associated with the lower susceptibility to disease and infection seen in the Fulani suggest a protective role of these antibodies in the Fulani. Large numbers of seroepidemiological studies in different malaria endemic areas have demonstrated an association between cytophilic subclasses IgG3 and/or IgG1 and protection against *P. falciparum* malaria.[88-90] No significant differences in the geometric mean concentrations of neither IgG2 nor IgG4 antibodies were detected in Fulani individuals of Burkina Faso and Mali as compared to their sympatric neighbours, Mossi and Dogon, respectively.

IgG antibody levels to most of other viral (measles), bacterial (*Helicobacter pylori; Mycobacterium tuberculosis*) and parasitic (*Toxoplasma gondii*) antigens tested did not differ between the different tribes, suggesting that the Fulani are not generally hyper-reactive.[87]

Cytokine-Mediated Responses to Plasmodium falciparum

The Fulani had significantly higher anti-malaria IgG and IgE antibodies and higher proportion of malaria specific IL-4 and IFN-γ producing cells compared to the Dogon. The higher percentage of IL-4 production in the less susceptible, the Fulani, may explain the difference in the antibody responses observed between the two study groups. Indeed, IL-4 also regulates the differenciation of precursor T-helper cells into the Th2 subset that regulates humoral immune responses including specific antibody production. The Fulani were more polarized towards Th2 than sympatric ethnic groups living under similar malaria transmission as evidenced by more IL-4 and less IL-12 producing cells and higher serum levels of anti-malaria IgG and IgE antibodies.

IFN-γ-183T allele, known to increase gene transcription, was associated with a significant (p = 0.009) reduced risk of cerebral malaria in children hospitalized at a pediatric ward in Bamako, Mali.[91]

Perspectives for Malaria Control: The Malaria Vaccine Approach

A malaria vaccine appears as a promising tool that will significantly boost the malaria control perspectives. Historically, vaccines have been the most cost-effective tools in the control of infectious diseases. Vaccination led to the eradication of smallpox from earth and to the elimination of poliomyelitis as a public health problem.

If we have evidence that a malaria vaccine is feasible, efforts to reach a vaccine that would be an effective public tool have not yet succeeded.[92]

Approaches to Design a Malaria Vaccine

P. falciparum is a complex organism. Its genome is constituted of 30 Mb while others pathogens for which a vaccine was relatively easily found such as the virus responsible for poliomyelitis has a genome size of 0.08 Mb. In addition, adaptation to its human ecological niche has lead to adaptive protein expression profiles and antigenic variation. The complexity of antigens expressed is additionally complicated by the allelic variation at loci. Furthermore, host genetic factors described earlier such as HLA type or hemoglobin type among the most studied restrict the human host immune response. A malaria vaccine may therefore have different efficacy and safety profile in different populations.

Despite the multiple constrains, significant progress have been made by malaria vaccinologists worldwide in the last 20 years toward a malaria vaccine. Today an unprecedented international effort is committed to malaria vaccine development. This commitment is being supported by developed and developing world leaders from the field of politics, science, pharmaceutical industry, economics and finances.

Malaria vaccinologists have adopted two principal approaches to design a malaria vaccine. Identification of the mechanisms of natural immunity and imitation of Mother Nature has constituted a productive approach. This approach led to the description of the antibody dependent cellular inhibition (ADCI), the model that best explains the status of semi immunity observed in population exposed to natural falciparum infection and protected against clinical malaria.[93] According to this model, the antigenic targets for efficient immune response are conserved. Subsequently immune response to parasite proteins such as MSP3 and GLURP were identified and found to correlate in sero-epidemiology surveys to protection against clinical malaria. As a result synthetic peptides based on MSP-3 have been developed and tested in phase 1 clinical trials in naïve and malaria-experienced adults living in Burkina Faso. Others synthetic peptides that combines MSP-3 and GLURP B-cell epitopes are being developed and will further enter the stage of field clinical evaluation.

The imitation of natural immunity also led to the discovery of parasite antigens critical for adhesion such as the *P. falciparum* erythrocyte membrane protein 1 (PfEMP1) that mediate parasite adhesion to several host receptors including host CD36, and chondroitin sulfate A (CSA) receptors. PfEMP1 is encoded by the large and diverse var gene family that is involved in clonal antigenic variation and plays a central role in *P. falciparum* pathogenesis. Exposure to pathogenic forms of *P. falciparum* harboring a variant PfEMP1 has induced protection against these parasites, leading to the selection of possibly less virulent parasites in subsequent infections.[94] Specific antibodies blocking parasite adhesion to CSA have been found to protect against placental infection. Hence, the functional restriction of highly variant parasite antigens has served to elaborate monoclonal antibodies targeting the CSA and aimed at protecting pregnant women against malaria.[95]

The second main approach taken by malaria vaccinologists was to assume that it is possible to achieve a level of protection better than what is observed in population naturally exposed to malaria. This approach focused to identify specific proteins on critical point of the parasite life cycle and to neutralize those proteins inducing specific strong cellular or antibodies responses. Malaria vaccine candidates have been produced based on various proteins expressed by sprorozoites (circum sporozoite (CS)-based constructs), at the surface of infected hepatocytes (liver stage antigen (LSA)-based constructs), at the surface of the infected red blood cells (merozoites surface proteins-MSP1, 2, 3, 4; Apical membrane antigens (AMA1), Erythrocyte binding antigen (EBA), glutamate rich protein (GLURP)), and proteins expressed during the sexual stage of the parasite development in the mosquito gut such as *P. falciparum* gamete surface antigens (Pfs230, Pfs48/45 and Pfs25) and ookinete surface antigen (Pfs28), and *Plasmodium vivax* ookinete surface antigen (Pvs25), for the transmission blocking strategy. Tremendous efforts have been achieved in the last ten years in the identification of candidate antigens, production of recombinant proteins, optimization of expression systems, and development of immune assays that correlates with protection against clinical disease.

Achievements toward a Malaria Vaccine

The networking among malaria researchers and the increased funding available for malaria research will foster the progress in the field of preclinical development of malaria vaccine candidates. The clinical testing of promising candidates is becoming a more and more acute need.

Optimal pathway from antigen discovery to vaccine delivered to target population includes the clinical development of the malaria vaccine candidates, once they have passed all steps of the preclinical characterization. The clinical development for a vaccine is a logical sequence of clinical trials in humans with the aims to establish that the malaria vaccine candidate is safe and it can protect against malarial disease. Until recently this step has been a major hurdle in the development of malaria vaccine for several reasons. Among the most important causes, are the high cost of the clinical trials, the need to perform the trials in population naturally exposed to malaria, the absence of endogen clinical trials capacity in most Sub-Saharan African countries, the diversity of the malaria infection requiring to test a malaria vaccine candidate in different populations exposed to different intensity of transmission, the ethical concerns of conducting clinical trials in vulnerable populations or in populations not exposed to malaria, the regulatory weaknesses in malaria endemic countries.

Despite these issues, several antigens have been evaluated in field clinical trials. The circumsporozoite (CS)-based construct, RTS,S codeveloped by GlaxoSmithKline Biologicals and Walter Reed Army Institute of Research has shown a significant protective efficacy of 57.7% against severe malaria in Mozambican children.[96] This was indeed a promising result. However the short duration of protection induced warrants further investigations to improve the vaccine efficacy. In addition a recent review of the trial data raised the possibility that non specific adjuvant related immune responses targeting hepatocytes could have been involved in the protective efficacy observed.[97] Association of the same adjuvant with others antigens candidate malaria vaccine will allow testing this hypothesis. Others malaria vaccine candidates have shown interesting immune responses and safety profile. That was the case of the MSP1 and AMA1-derived recombinant antigens.[98]

An Alternative for the 21st Century

New adjuvant systems such as the CpG ODN hold the promise of inducing strong cellular immune responses,[99] although, this potential need to be confirmed in population naturally exposed to malaria. Recently, Mueller et al, have induced sterile protection in the rodent model, using genetically-modified sporozoites that were unable to develop in the host liver cells.[100] Such use of the genomic tools opens an extremely promising way to research. Immunizations with irradiated sporozoite have until today produce the best protective immune responses in all malaria animal models. With a complex organism such as *P. falciparum*, the use of whole parasite may be the best strategy to induce an immune response covering enough antigen repertoires to efficiently prevent malaria disease. Most malaria vaccinologists believe that the effective malaria vaccine will be a construct of different epitopes from different parasite proteins covering the stage-specific proteins and their variants.

There is still the need to accelerate the pace of the malaria vaccine research. Given the limitations of existing animal models[101,102] and the difficulties related to the impact of any interventions against malaria, there is only one way to validate efficacy of malaria vaccine candidate: testing them in clinical trials in endemic countries. We believe that a critical part of the international mobilization against malaria should target the development of clinical trials capacity and overall biomedical research capacity in Africa. Only few sites, located in Gambia, Mozambique, Kenya, Tanzania and South Africa are presently capable of conducting large efficacy clinical trials under international standards. In others countries such as Mali, Burkina Faso, Ghana, the capacity to conduct phase I and II clinical trials is being upgraded. This is still insufficient to face the clinical development need of all antigens in the pipeline. The challenge study model[103] was developed as an efficient model to test the efficacy of malaria vaccine candidate targeting preerythrocytic stage antigens. There are several confounders that may affect the efficacy of the challenge model in malaria endemic countries or assessing a vaccine targeting blood stage antigens. Most of the confounders can be controlled by a rigorous study design. And carefully designed molecular tools can help discriminating persistent from new infections parasites. We believe that the challenge studies should be conducted in malaria endemic countries, preferably in areas of seasonal transmission.

Conclusion

The grand challenge for us is to develop strategies to identify gene variants that contribute to good health and resistance to severe malaria as our ultimate goal is the reduction of morbidity and mortality due to malaria through prevention and treatment. The next steps for scientists working in this area will be to authenticate known associations, to understand the functional basis of genetic associations, to screen for entirely novel malaria resistance genes and to investigate gene-environment interactions. Genomics will facilitate further understanding of this aspect of human biology and allow the identification of gene variants that are important for the maintenance of health, particularly in the presence of known environmental risk factors.[104] Another possible approach is the full examination of genetic variants in children at high risk (under 5) for severe malaria and, who do not develop them. Such strategies should enable the research community to achieve the following: (1) identify human genes (and genes products) and pathways with a role in health and malaria disease, and determine how they interact with environment factors; (2) Develop, evaluate and apply genome-based diagnostic methods for the prediction of susceptibility to malaria.

Finally, training and maintaining of African capacities in genomics will be the best investment for the purpose of testing and validating genomic based technologies for public health use in Africa.

History of malaria eradication and control tell us that the disease will be rolled back from Africa only when technological advances will be applied together with social and economical development, with the African people, their leaders and their scientists involved as key actors of the battle (Box 1). And we believe that such involvement should derive from a profound inner

---

Box 1.   Transgenic mosquitoes

An important line of ongoing research is to develop transgenic mosquitoes, i.e., mosquitoes genetically manipulated, so that they can no longer transmit malaria and use these mosquitoes to replace the natural population. Although significant progress is being made (with identification of the mosquito genome and the progress in the biotechnology), it is not sure that this will be applied widely in a near future. More research is needed, specifically large field studies under different malaria transmission conditions and different ecological areas.

Box 2.   Technological advances, socio-economical development and malaria

---

History of malaria eradication and control tell us that the disease will be rolled back from Africa only when technological advances will be applied together with social and economical development, with the African people, their leaders and their scientists involved as key actors in the battle. And we believe that such involvement should derive from a profound inner movement and not dictated by non-African organizations. As discussed in this book by Castro and Millet,[105] we are convinced that macro-economic adjustments need to be reassessed and priority given to a real chance for development in Africa.

---

movement and not dictated by non African organizations. As discussed earlier in this book by Castro and Millet (Box 2),[105] we are convinced that macro economic adjustments need to be reassessed and priority given to a real chance for development in Africa.

Acknowledgements

This work was supported by the following funding agencies: NIAID/NIH (National Institute of Allergy and Infectious Diseases/National Institute of Health); CVD - University of Maryland (Center for Vaccine Development - University of Maryland); TDR/WHO (Tropical Disease Research/World Health Organization); TDR/MIM (Tropical Disease Research/Multilateral Initiative on Malaria); and AUF (Agence Universitaire Francophone).

References

1. Beltran E. Commemoration of the discovery of the pathogenic germs of malaria. I. Charles Louis Alphonse Laveran (1845-1922). His 1880 discovery. Gac Med Mex 1981; 117:195-201.
2. Ward RA. The influence of Ronald Ross upon the early development of malaria vector control procedures in the United States Army. J Trop Med Hyg 1973; 76:207-209.
3. Ross R. The role of the mosquito in the evolution of the malarial parasite: The recent researches of Surgeon-Major Ronald Ross, I.M.S. 1898. Yale J Biol Med 2002; 75:103-105.
4. Bruce-Chwatt LJ. Essential Malariology. 2 ed. New York: John Wiley & Sons, 1985.
5. Bruce-Chwatt LJ. Essential Malariology. 2 ed. New York: Oxford University Press, Inc., 1993.
6. Peters W. Chemotherapy and drug resistance in malaria. 2 ed. London: Academic Press, 1987.
7. Meshnick SR, Dobson MJ. The history of antimalarial drugs. In: Rosenthal PJ, ed. Antimalarial Chemotherapy. Mechanisms of Action, Resistance and New Direction in Drug Discovery. Totowa, New Jersey: Humana Press, 2001:15-25.
8. Olliaro P, Cattani J, Wirth D. Malaria, the submerged disease. JAMA 1996; 275:230-233, Review.
9. WHO. Report of the malaria conference in Equatorial Africa, Kampala. Technical Report Series 38,. Geneva; World Health Organization, 1951.
10. Metselaar D, van Thiel PM. Classification of malaria. Tropical and Geographical Medicine 1959; 11:157-161.
11. Molineaux L. The Epidemiology of human malaria as an explanation of distribution, including some implications for its control. In: Wensdorfer WH, Mc Gregor I, eds. Malaria Principles and Practices of Malariology, 2nd ed. Edingburgh, London, New York: 1988.
12. Clarke SE, Brooker S, Njagi JK et al. Malaria morbidity among school children living in two areas of contrasting transmission in western Kenya. Am J Trop Med Hyg 2004; 71:732-738.
13. Snow RW, Omumbo JA, Lowe B et al. Relation between severa malaria morbidity in children and level of Plasmodium falciparum transmission in Africa. Lancet 1997; 349 (9066):1650-1654.
14. Snow RW, Marsh K. The consequences of reducing transmission of Plasmodium falciparum in Africa. Adv Parasitol 2002; 52:235-264.
15. Trape JF, Rogier C. Combating malaria morbidity and mortality by reducing malaria transmission. Parasitol Today 1996; 12:236-240.
16. World Health Organization. Severe Malaria. Trans R Soc Trop Med Hyg 2000; 94(Suppl 1):1-90.
17. Harrison NE, Odunukwe NN, Ijoma CK et al. Current clinical presentation of malaria in Enugu. Niger Postgrad Med J 2004; 11:240-245.
18. Loveridge BW, Henner JR, Lee FC. Accurate clinical diagnosis of malaria in a postflood epidemic: A field study in Mozambique. Wilderness Environ Med 2003; 14:17-19.
19. Robinson P, Jenney AW, Tachado M et al. Imported malaria treated in Melbourne, Australia: Epidemiology and clinical features in 246 patients. J Travel Med 2001; 8:76-81.

20. Armstrong-Schellenberg JRM, Smith T, Alonso PL et al. What is clinical malaria? Finding case definitions for field research in highly endemic areas. Parasitol Today 1994; 10:439-442.

21. Smith T, Killeen G, Lengeler C et al. Relationships between the outcome of Plasmodium falciparum infection and the intensity of transmission in Africa. Am J Trop Med Hyg 2004; 71:80-86.

22. Bloland PB, Boriga DA, Ruebush TK et al. Longitudinal cohort study of the epidemiology of malaria infections in an area of intense malaria transmission II. Descriptive epidemiology of malaria infection and disease among children. Am J Trop Med Hyg 1999; 60:641-648.

23. Dicko A, Mantel C, Kouriba B et al. Season, fever prevalence and pyrogenic threshold for malaria disease definition in an endemic area of Mali. Trop Med Int Health 2005; 10:550-556.

24. Rogier C, Commenges D, Trape JF. Evidence for an age-dependent pyrogenic threshold of Plasmodium falciparum parasitemia in highly endemic populations. Am J Trop Med Hyg 1996; 54:613-619.

25. Vounatsou P, Smith T, Kitua AY et al. Apparent tolerance of Plasmodium falciparum in infants in a highly endemic area. Parasitology 2000; 120(Pt 1):1-9.

26. Dicko A, Klion AD, Thera MA et al. The etiology of severe anemia in a village and a periurban area in Mali. Blood 2004; 104:1198-1200.

27. Snow RW, Bastos DA I, Lowe BS et al. Severe childhood malaria in two areas of markedly different falciparum transmission in east Africa. Acta Trop 1994; 57:289-300.

28. Roll Back Malaria. Framework for monitoring progress and evaluating outcomes & impact; Strategic framework for malaria control during pregnancy in the Africa Region. World Health Organization 2002:1-6.

29. Shulman CE, Dorman EK. Importance and prevention of malaria in pregnancy. Trans R Soc Trop Med Hyg 2003; 97:30-35.

30. Steketee RW, Wirima JJ, Campbell CC. Developing effective strategies for malaria prevention programs for pregnant African women. Am J Trop Med Hyg 1996; 55:95-100.

31. McDermott JM, Wirima JJ, Steketee RW et al. The effect of placental malaria infection on perinatal mortality in rural Malawi. Am J Trop Med Hyg 1996; 55:61-65.

32. Gysin J, Pouvelle B, Fievet N et al. Ex vivo desequestration of Plasmodium falciparum-infected erythrocytes from human placenta by chondroitin sulfate A. Infect Immun 1999; 67:6596-6602.

33. McCormick MC. The contribution of low birth weight to infant mortality and childhood morbidity. N Engl J Med 1985; 312:82-90.

34. Steketee RW, Wirima JJ, Slutsker L et al. The problem of malaria and malaria control in pregnancy in sub-Saharan Africa. Am J Trop Med Hyg 1996; 55:2-7.

35. Breman JG, Alilio MS, Mills A. Conquering the intolerable burden of malaria: What's new, what's needed: A summary. Am J Trop Med Hyg 2004; 71:1-15.

36. Snow RW, Craig M, Deichmann U et al. Estimating mortality, morbidity and disability due to malaria among Africa's nonpregnant population. Bull World Health Organ 1999; 77:624-640.

37. Snow RW, Korenromp EL, Gouws E. Pediatric mortality in Africa: Plasmodium falciparum malaria as a cause or risk? Am J Trop Med Hyg 2004; 71:16-24.

38. Carneiro I, Roca-Feltrer A, Schellenberg J. Estimate of the burden of malaria morbidity in Africa in children under the age of five years; Child Helath Epidemiology Reference Group Working Paper. London School of Hygiene and Tropical Medicine 2005.

39. Rowe AK, Steketee RW, Snow RW et al. Estimates of the burden of the mortality directly attributable to malaria for children under 5 years of age in Africa for the year 2000. Geneva: World Health Organization, 2004.

40. Kager PA. Malaria control: Constraints and opportunities. Trop Med Int Health 2002; 7:1042-1046.

41. UNICEF. The state of the world's children. Oxford: Oxford University Press, 1998.

42. World Health Organization. A global strategy for malaria. Geneva, Switzerland: World Health Organization, 1993.

43. Trape JF. The public health impact of chloroquine resistance in Africa. Am J Trop Med Hyg 2001; 64:12-17.

44. Djimde A, Doumbo OK, Cortese JF et al. A molecular marker for chloroquine-resistant falciparum malaria. N Engl J Med 2001; 344:257-263.

45. Jelinek T, Aida AO, Peyerl-Hoffmann G et al. Diagnostic value of molecular markers in chloroquine-resistant falciparum malaria in Southern Mauritania. Am J Trop Med Hyg 2002; 67:449-453.

46. Kublin JG, Dzinjalamala FK, Kamwendo DD et al. Molecular markers for failure of sulfadoxine-pyrimethamine and chlorproguanil-dapsone treatment of Plasmodium falciparum malaria. J Infect Dis 2002; 185:380-388.

47. Bwijo B, Kaneko A, Takechi M et al. High prevalence of quintuple mutant dhps/dhfr genes in Plasmodium falciparum infections seven years after introduction of sulfadoxine and pyrimethamine as first line treatment in Malawi. Acta Trop 2003; 85:363-373.

48. Kyabayinze D, Cattamanchi A, Kamya MR et al. Validation of a simplified method for using molecular markers to predict sulfadoxine-pyrimethamine treatment failure in African children with falciparum malaria. Am J Trop Med Hyg 2003; 69:247-252.

49. Tinto H, Ouedraogo JB, Erhart A et al. Relationship between the Pfcrt T76 and the Pfmdr-1 Y86 mutations in Plasmodium falciparum and in vitro/in vivo chloroquine resistance in Burkina Faso, West Africa. Infect Genet Evol 2003; 3:287-292.

50. McIntosh HM, Olliaro P. Artemisinin derivatives for treating severe malaria. Cochrane Database Syst Rev 2000; CD000527.

51. Mutabingwa TK, Anthony D, Heller A et al. Amodiaquine alone, amodiaquine+sulfadoxine-pyrimethamine, amodiaquine+artesunate, and artemether-lumefantrine for outpatient treatment of malaria in Tanzanian children: A four-arm randomised effectiveness trial. Lancet 2005; 365:1474-1480.

52. Omari AA, Gamble C, Garner P. Artemether-lumefantrine for uncomplicated malaria: A systematic review. Trop Med Int Health 2004; 9:192-199.

53. Drakeley CJ, Jawara M, Targett GA et al. Addition of artesunate to chloroquine for treatment of Plasmodium falciparum malaria in Gambian children causes a significant but short-lived reduction in infectiousness for mosquitoes. Trop Med Int Health 2004; 9:53-61.

54. von SL, Jawara M, Coleman R et al. Parasitaemia and gametocytaemia after treatment with chloroquine, pyrimethamine/sulfadoxine, and pyrimethamine/sulfadoxine combined with artesunate in young Gambians with uncomplicated malaria. Trop Med Int Health 2001; 6:92-98.

55. Challis K, Osman NB, Cotiro M et al. Impact of a double dose of sulphadoxine-pyrimethamine to reduce prevalence of pregnancy malaria in southern Mozambique. Trop Med Int Health 2004; 9:1066-1073.

56. Kayentao K, Kodio M, Newman RD et al. Comparison of intermittent preventive treatment with chemoprophylaxis for the prevention of malaria during pregnancy in Mali. J Infect Dis 2005; 191:109-116.

57. Mubyazi G, Bloch P, Kamugisha M et al. Intermittent preventive treatment of malaria during pregnancy: A qualitative study of knowledge, attitudes and practices of district health managers, antenatal care staff and pregnant women in Korogwe District, North-Eastern Tanzania. Malar J 2005; 4:31.

58. Massaga JJ, Kitua AY, Lemnge MM et al. Effect of intermittent treatment with amodiaquine on anaemia and malarial fevers in infants in Tanzania: A randomised placebo-controlled trial. Lancet 2003; 361:1853-1860.

59. Schellenberg D, Menendez C, Kahigwa E et al. Intermittent treatment for malaria and anaemia control at time of routine vaccinations in Tanzanian infants: A randomised, placebo-controlled trial. Lancet 2001; 357:1471-1477.

60. O'Meara WP, Breman JG, McKenzie FE. Intermittent preventive malaria treatment in Tanzanian infants. Lancet 2005; 366:545-546.

61. Tishkoff SA, Williams SM. Genetic analysis of African populations: Human evolution and complex disease. Nat Rev Genet 2002; 3:611-621.

62. Dolo A, Modiano D, Maiga B et al. Difference in susceptibility to malaria between two sympatric ethnic groups in Mali. Am J Trop Med Hyg 2005; 72:243-248.

63. Modiano D, Petrarca V, Sirima BS et al. Different response to Plasmodium falciparum malaria in west African sympatric ethnic groups. Proc Natl Acad Sci USA 1996; 93:13206-13211.

64. Hill AV. The genomics and genetics of human infectious disease susceptibility. Annu Rev Genomics Hum Genet 2001; 2:373-400.

65. Kwiatkowski DP. How malaria has affected the human genome and what human genetics can teach us about malaria. Am J Hum Genet 2005; 77:171-190.

66. Hill AV, Allsopp CE, Kwiatkowski D et al. Common west African HLA antigens are associated with protection from severe malaria. Nature 1991; 352:595-600.

67. Foote SJ, Burt RA, Baldwin TM et al. Mouse loci for malaria-induced mortality and the control of parasitaemia. Nat Genet 1997; 17:380-381.

68. Miller LH. Impact of malaria on genetic polymorphism and genetic diseases in Africans and African Americans. Proc Natl Acad Sci USA 1994; 91:2415-2419.

69. Weatherall DJ. Common genetic disorders of the red cell and the 'malaria hypothesis'. Ann Trop Med Parasitol 1987; 81:539-548.

70. Hereditary anaemias: Genetic basis, clinical features, diagnosis, and treatment. WHO working group. Bull World Health Organ 1982; 60:643-660.
71. Flint J, Hill AV, Bowden DK et al. High frequencies of alpha-thalassaemia are the result of natural selection by malaria. Nature 1986; 321:744-750.
72. Fleming AF, Storey J, Molineaux L et al. Abnormal haemoglobins in the Sudan savanna of Nigeria. I. Prevalence of haemoglobins and relationships between sickle cell trait, malaria and survival. Ann Trop Med Parasitol 1979; 73:161-172.
73. Modiano D, Luoni G, Sirima BS et al. Haemoglobin C protects against clinical Plasmodium falciparum malaria. Nature 2001; 414:305-308.
74. Rihet P, Flori L, Traore AS et al. Hemoglobin C is associated with reduced Plasmodium falciparum parasitemia and low risk of mild malaria attack. Hum Mol Genet 2004; 13:1-6.
75. Agarwal A, Guindo A, Cissoko Y et al. Hemoglobin C associated with protection from severe malaria in the Dogon of Mali, a West African population with a low prevalence of hemoglobin S. Blood 2000; 96:2358-2363.
76. Fairhurst RM, Baruch DI, Brittain NJ et al. Abnormal display of PfEMP-1 on erythrocytes carrying haemoglobin C may protect against malaria. Nature 2005; 435:1117-1121.
77. Ruwende C, Khoo SC, Snow RW et al. Natural selection of hemi- and heterozygotes for G6PD deficiency in Africa by resistance to severe malaria. Nature 1995; 376:246-249.
78. Piazza A, Mayr WR, Contu L et al. Genetic and population structure of four Sardinian villages. Ann Hum Genet 1985; 49(Pt 1):47-63.
79. McGuire W, Hill AV, Allsopp CE et al. Variation in the TNF-alpha promoter region associated with susceptibility to cerebral malaria. Nature 1994; 371:508-510.
80. Knight JC, Udalova I, Hill AV et al. A polymorphism that affects OCT-1 binding to the TNF promoter region is associated with severe malaria. Nat Genet 1999; 22:145-150.
81. Urban BC, Ferguson DJ, Pain A et al. Plasmodium falciparum-infected erythrocytes modulate the maturation of dendritic cells. Nature 1999; 400:73-77.
82. McGilvray ID, Serghides L, Kapus A et al. Nonopsonic monocyte/macrophage phagocytosis of Plasmodium falciparum-parasitized erythrocytes: A role for CD36 in malarial clearance. Blood 2000; 96:3231-3240.
83. Serghides L, Smith TG, Patel SN et al. CD36 and malaria: Friends or foes? Trends Parasitol 2003; 19:461-469.
84. Aitman TJ, Cooper LD, Norsworthy PJ et al. Malaria susceptibility and CD36 mutation. Nature 2000; 405:1015-1016.
85. Pain A, Urban BC, Kai O et al. A nonsense mutation in Cd36 gene is associated with protection from severe malaria. Lancet 2001; 357:1502-1503.
86. Omi K, Ohashi J, Naka I et al. Polymorphisms of CD36 in Thai malaria patients. Southeast Asian J Trop Med Public Health 2002; 33(Suppl 3):1-4.
87. Bolad A, Farouk SE, Israelsson E et al. Distinct interethnic differences in immunoglobulin G class/subclass and immunoglobulin M antibody responses to malaria antigens but not in immunoglobulin G responses to nonmalarial antigens in sympatric tribes living in West Africa. Scand J Immunol 2005; 61:380-386.
88. Aribot G, Rogier C, Sarthou JL et al. Pattern of immunoglobulin isotype response to Plasmodium falciparum blood-stage antigens in individuals living in a holoendemic area of Senegal (Dielmo, west Africa). Am J Trop Med Hyg 1996; 54:449-457.
89. Chumpitazi BF, Lepers JP, Simon J et al. IgG1 and IgG2 antibody responses to Plasmodium falciparum exoantigens correlate inversely and positively, respectively, to the number of malaria attacks. FEMS Immunol Med Microbiol 1996; 14:151-158.
90. Taylor RR, Allen SJ, Greenwood BM et al. IgG3 antibodies to Plasmodium falciparum merozoite surface protein 2 (MSP2): Increasing prevalence with age and association with clinical immunity to malaria. Am J Trop Med Hyg 1998; 58:406-413.
91. Cabantous S, Poudiougou B, Traore A et al. Evidence that interferon- gamma plays a protective role during cerebral malaria. J Infect Dis 2005; 192:854-860.
92. Moorthy VS, Good MF, Hill AV. Malaria vaccine developments. Lancet 2004; 363:150-156.
93. Bouharoun-Tayoun H, Oeuvray C, Lunel F et al. Mechanisms underlying the monocyte-mediated antibody-dependent killing of Plasmodium falciparum asexual blood stages. J Exp Med 1995; 182:409-418.
94. Miller LH, Baruch DI, Marsh K et al. The pathogenic basis of malaria. Nature 2002; 415:673-679.
95. Lekana Douki JB, Traore B, Costa FT et al. Sequestration of Plasmodium falciparum-infected erythrocytes to chondroitin sulfate A, a receptor for maternal malaria: Monoclonal antibodies against the native parasite ligand reveal pan-reactive epitopes in placental isolates. Blood 2002; 100:1478-1483.

96. Alonso PL, Sacarlal J, Aponte JJ et al. Efficacy of the RTS,S/AS02A vaccine against Plasmodium falciparum infection and disease in young African children: Randomised controlled trial. Lancet 2004; 364:1411-1420.
97. Snounou G, Gruner AC, Muller-Graf CD et al. The Plasmodium sporozoite survives RTS,S vaccination. Trends Parasitol 2005.
98. Heppner Jr DG, Kester KE, Ockenhouse CF et al. Towards an RTS,S-based, multi-stage, multi-antigen vaccine against falciparum malaria: Progress at the Walter Reed Army Institute of Research. Vaccine 2005; 23:2243-2250.
99. Weeratna RD, Brazolot Millan CL, McCluskie MJ et al. CpG ODN can redirect the Th bias of established Th2 immune responses in adult and young mice. FEMS Immunol Med Microbiol 2001; 32:65-71.
100. Mueller AK, Labaied M, Kappe SH et al. Genetically modified Plasmodium parasites as a protective experimental malaria vaccine. Nature 2005; 433:164-167.
101. Heppner DG, Cummings JF, Ockenhouse C et al. New World monkey efficacy trials for malaria vaccine development: Critical path or detour? Trends Parasitol 2001; 17:419-425.
102. Stowers AW, Miller LH. Are trials in New World monkeys on the critical path for blood-stage malaria vaccine development? Trends Parasitol 2001; 17:415-419.
103. Church LW, Le TP, Bryan JP et al. Clinical manifestations of Plasmodium falciparum malaria experimentally induced by mosquito challenge. J Infect Dis 1997; 175:915-920.
104. Guttmacher AE, Collins FS. Genomic medicine—a primer. N Engl J Med 2002; 347:1512-1520.
105. Castro J, Millet D. Malaria and structural adjustment: Proof by contradiction. In: Boëte C, ed. Genetically Modified Mosquitoes for Malaria Control. Georgetown: Landes Bioscience, 2006; 2:16-23.

CHAPTER 2

Malaria and Structural Adjustment:
Proof by Contradiction

Julie Castro* and Damien Millet

Summary

The evolution of malaria over the last thirty years shows that far from regressing, the disease is actually in a process of reglobalisation. While health institutions confine themselves to analysing the economic impact of the pandemic, the authors call for a reposing of the problem in other terms. The unprecedented resurgence of malaria is in fact contemporary with the application of structural adjustment programmes, as devised and implemented by international financial institutions since the 1980s. An examination of these programmes reveals that they serve to organise and secure the transfer of wealth from populations in the South to the ruling classes of the South and North. Their considerable economic and social impact interferes with the determinants of malaria on several levels and explains the resurgence of the disease. Applying solely technical solutions to the problem is therefore not enough: to reverse the trend, it is the current world economic order that must radically change.

A Scourge in Full Resurgence

The figures on malaria speak volumes: every year, between 300 and 500 million people are affected by this disease, which claims between 1.7 and 2.5 million lives. More than 40% of the world population is currently exposed to malaria and the number of confirmed cases is on the increase. The regions concerned are primarily in tropical and sub-tropical zones, among which Sub-Saharan Africa pays the heaviest toll, with more than 90% of fatalities. Malaria's primary targets are the young: it is estimated that the disease kills an African child every thirty seconds and that it is the leading cause of child mortality in Sub-Saharan Africa.

The evolution of the disease over time is closely related to the history of humanity, and in this respect the broad epidemiological outlines of the last century are worth noting. At the beginning of the twentieth century, malaria affected a greater number of regions in the world, going on to become almost obsolete in temperate zones some fifty years ago. This geographical recession was accompanied by a significant fall-off in the number of cases: in 1950, malaria accounted for 1.2 million deaths, with numbers subsequently decreasing to an annual death rate of 500,000 in 1970. The reversal of this trend, as observed over the last twenty years, testifies to a resurgence of unprecedented proportions: malaria is gaining ground on every front, even reemerging in regions where it had been totally eradicated, such as the Middle East, Turkey and Central Asia.

At the same time, the last twenty years have seen the advent of a global neo-liberal economic model commonly known as globalisation. In the countries of the South and in Sub-Saharan

*Corresponding Author: Julie Castro—Comité pour l'Annulation de la Dette du Tiers-Monde (CADTM) France, 17 rue de la bate, 45150 Jargeau, France. Email: julie.castro@cadtm.org

Genetically Modified Mosquitoes for Malaria Control, edited by Christophe Boëte.
©2006 Landes Bioscience.

Africa in particular, strict macro-economic policies, going by the name of Structural Adjustment Programmes (SAPs), have been forcibly introduced by international financial institutions via the debt mechanism. Is the coexistence of these two developments a mere coincidence? To what degree does this new economic world order account for the reglobalisation of malaria?

A Restrictive Ideological Framework

The medicoscientific approach to malaria is the one most frequently adopted. This type of reasoning, supposedly neutral because it is scientific, nevertheless constitutes an ideological position in itself: by "naturalising" the disease, research into its causes and the creation of solutions are restricted to this single line of thinking. The same ideological stance can also be observed at the institutional level. Take, for example, the "*Roll Back Malaria*" campaign - an initiative of the World Health Organisation (WHO), the United Nations Children's Fund (UNICEF), the United Nations Development Programme (UNDP) and the World Bank. In an introductory document[1] on its website, we find a succinct description of the different parasites, and of the cycles in host and vector. The existence of nonscientific determining factors for the disease, and in particular those of a socio-economic nature, is totally avoided.

The consequences of malaria, however, are approached first from an economic angle, and only secondly in human terms. This predominantly economic standpoint is assumed without compunction. And while a "*Roll Back Malaria*" document concedes that "*in Africa today, malaria is understood to be both a disease of poverty and a cause of poverty*"[2] it then goes on to say: "*Annual economic growth in countries with high malaria transmission has historically been lower than in countries without malaria. Economists believe that malaria is responsible for a growth penalty of up to 1.3% per year in some African countries*". The authors seem almost to regret that the human consequences cannot be more easily translated into figures when they write: "*Malaria has a greater impact on Africa's human resources than simple lost earnings. Although difficult to express in dollar terms, another indirect cost of malaria is the human pain and suffering caused by the disease*".[3]

The direct and indirect costs of the disease are moreover the subject of in-depth, detailed analyses: "*In some countries with a heavy malaria burden, the disease may account for as much as 40% of public health expenditure, 30-50% of inpatient admissions, and up to 50% of outpatient visits*".[2] And the assessment of the economic impact of malaria by the WHO is nothing short of vertiginous: "*As a result of the cumulative effect over thirty-five years, the gross domestic product of African countries is today 32% lower that it would have been without malaria*".[4] In terms of the family, this means that a poor African family can find itself spending a quarter of its annual income on the prevention and treatment of malaria.[5]

The devastating impact of the pandemic in economic terms does not preclude a critical examination of the ideological process. The present analysis is utilitarian: it fails to challenge the responsibility of economic policies in the entrenched presence or resurgence of diseases, and restricts itself to assessing the consequences. In this context, improving public health is a mean to improving economic health, and not the opposite. This ethically unacceptable view of health as an instrument of economic profitability is carried through to a number of other organisations to encompass the whole health sector. For example, the WHO, whose public aid policy mission, as stated in 1980, was to ensure "health for all", took an entirely different turn under the presidency of Dr. Gro Harlem Brundtland. She opened the field to the private sector, the World Bank, the IMF and the WTO in the late 1990s at the 51st World Health Assembly, and her policies are now governed by free economy rules.[6] This shift in the conception of health—as a means to production and no longer a right—was clearly expressed by Dr. Bruntland in 2000 during the third International Conference on Priorities in Health Care in Amsterdam. "Improvement in health will significantly increase the forces for economic development and the reduction of poverty".[7] Finally, the creation in 2001 of the Global Health Fund, an independent private foundation focussing on AIDS, malaria and tuberculosis, seems to indicate that global public health issues will increasing elude the domain of the WHO, which sits on the board but has no vote.

In the face of a situation where health problems are considered upstream of the economic situation, it is essential to rethink the problem: malaria is as much a social and political process as a scientific reality. As such, one must first assess the impact of macro-economic neo-liberal policies on the evolution of the pandemic. However, in the interests of preserving the foundations of the prevailing economic model, no such assessment is being made, and is even being deliberately ignored. One of the nerve centres of this model, which, as we will see, bears a heavy responsibility in the gravity of the pandemic, is debt. It is debt that has allowed creditors, both public and private, to bring a Trojan horse into the economy of the South: its name is structural adjustment, and behind this rather creative designation lies a very efficient system of domination.

A Ruthless Tool of Domination

At the beginning of the 1960s, when the countries of Latin America were fast industrialising, the nations of Asia, followed by Africa, gained their independence. In the space of twenty years they were to shift from a colonial style of political domination to a system of dominance less visible, yet no less effective, via the debt mechanism. This debt was initially encouraged by private creditors in the North, whose coffers were overflowing with liquid assets (eurodollars, then petrodollars after the oil crisis of 1973). The leaders of the wealthy countries were next to come on the scene. In the 1970s, to revive their failing economies, they offered the developing countries loans that were conditioned by the purchase of goods from the creditor country: a system called linked aid. Finally, on the multilateral front, the World Bank was to launch, under the presidency of Robert Mc Namara, an aggressive policy of inducement to borrow: from 1968, when Mc Namara took office, to 1973, the World Bank lent more than at any time since 1945. It used debt for geo-strategic purposes, whether to strengthen the Western bloc's strategic alliances (Suharto in Indonesia, Marcos in the Philippines, Mobutu in Zaire, Pinochet in Chile, the dictator states of Argentina and Brazil, etc.) or to ensure that countries attracted by a nationalistic and economically independent state system stay within the fold of the U.S. Debt became a key tool in the domination of Third World countries.

A decisive turning point came in the 1980s. A drastic rise in interest rates, decided unilaterally by the United States, and then by the United Kingdom in the late 1970s, vastly increased the sums to be refunded. In addition, a glut on the commodities market of the very raw materials exported by the countries in the South caused prices to plummet. The production of agricultural and mining industry commodities, most often exported unprocessed, had been vigorously encouraged by the major powers, who were the main consumers. For the countries of the South, these exports were the principal means of obtaining the foreign currency they needed to repay their debt. The wheels were in motion: developing countries must export more and yet more to earn the same amount. So began a relentless spiral governed by a policy of "everything for export": revenues soon dropped dramatically. These same countries also had a harder time finding find lenders because investors were giving preference to the wealthy countries (offering higher interest rates) where there were juicy profits to be made. Tripled interest rates, diminishing revenues, a scarcity of new loans: the situation soon became intolerable. In 1982, the crisis reached a head. Mexico was the first to announce it could no longer meet its engagements.

The creditors, alarmed by the spate of repayment cessation notices, mandated the International Monetary Fund (IMF) to organise and secure repayment of the debt. As an immediate measure, the IMF authorised new loans to developing countries so as to avoid default wherever possible. In exchange, developing countries were forced to apply economic policies dictated by the financial experts, the so-called "structural adjustment programmes", whose explicit purpose was to increase the resources of indebted countries to enable them to repay the debt. The means adopted to achieve this objective were to have social, economic and human consequences of catastrophic proportions.

Servicing the Debt at All Costs

The recommended measures inherent in a Structural Adjustment Programme always follow the same pattern:

- Stopping of subsidies for basic products and services (bread, rice, milk, sugar, fuel, etc.). To compensate for the lack of a guaranteed minimum income, governments normally step in to keep basic foods as well as other vital goods and services at affordable prices. In a SAP, the IMF requires that this form of subsidy be discontinued. With harsh consequences for the population: the price of basic foods and fuel skyrockets. Not only do higher fuel prices aggravate the water problem (because boiling water to make it potable costs more), but they also lead to an increase in the cost of transportation, which in its turn affects the price of foodstuffs sold in the local market. Thus these measures systematically increase food insecurity, lead to inflationary prices and weaken the local economy.
- Severe budgetary measures and reduction of public spending, generally via drastic cuts in allegedly "nonproductive" social budgets (health, education, housing, infrastructures).
- Devaluation of the local currency, with a view to making local exported products cheaper and thus more competitive on the world market. In theory, this makes it easier for them to find buyers. The downside is that more products must be sold simply to get back the same amount of foreign currency as before. But taking into account that several countries will be following this procedure at the same time, the logic falls apart, since the simultaneous arrival of identical products on the market causes prices to fall. Conversely, foreign products are more expensive locally (such as mosquito nets imported to help fight malaria). To take an example: in January 1994, the IMF and France got the African governments in question to devalue the CFA Franc by 50%. The results were disastrous: a product imported from France which formerly cost 100 CFA Francs went up to 200 CFA Francs overnight, and for an African country to get back 100 FF it now meant selling double the amount of products. The purchasing power of populations in the CFA zone took a steep dive, even more so since wages were frozen. At the same time, the debt for these countries (payable in hard currency) was suddenly doubled. So that they had to earn twice the amount (in local currency) to get the foreign currency they needed to pay back the debt.
- Higher interest rates, to attract foreign capital with the prospect of high returns. For small farmers borrowing on the local market to buy seed, fertiliser and agricultural machinery, this rise in interest rates drastically reduces their borrowing power: sowing is curtailed and agricultural production falls off. Indebted companies have to face up to unforeseen repayments in an already depressed market - a situation that leads to numerous bankruptcies. Lastly, the hike in interest rates increases the burden of internal public debt for the government, hence a larger public deficit, when paradoxically the aim of SAPs is to reduce it. The government then proceeds to make even more drastic cuts in social spending, creating a truly vicious circle.
- Agricultural production entirely focussed on export (coffee, cotton, cocoa, groundnuts, tea, etc.) to bring in currency. This "everything for export" has as its corollary a decline in the production of basic food crops (cassava or millet for example) and increased deforestation. The countries in question are led to specialise in one (or several) agricultural products for export, one (or several) mineral products or primary industries such as fishing, thereby becoming dependent on this particular resource or crop. Economic instability increases, since prices on the world market can suddenly suffer a huge drop, causing the collapse of a country's whole economy. For the environment too, this relentless race for productivity has catastrophic consequences: deforestation, threatened biodiversity, systematic use of insecticides and fertilisers, soil erosion and impoverishment, etc.
- Total opening up of markets by the elimination of tariff barriers, liberalisation of trade and capital markets, and the removal of exchange controls. To attract foreign investors, the arrival of foreign multinationals is encouraged. Thanks to their financial weight and superior technology, they grab significant market share in numerous business sectors to the

detriment of many local producers who are forced into liquidation. Another result of trade liberalisation is the removal of any obstacles that would prevent multinationals rechannelling profits generated locally to their headquarters in the North. The wealth produced in the South is thus siphoned off to benefit the wealthy shareholders of these corporations. Meanwhile, inflation and growing unemployment wreak havoc among the local population.

• A tax system creating even greater inequality thanks to the principle of a value added tax (VAT) and the preservation of capital earnings. The removal of tariff barriers reduces tax revenues, resulting in wider-scale taxation that penalises the poor in particular, with the suppression of progressive taxation and a generalised VAT system. Take the case of a VAT of 18% as in French-speaking West Africa. It is applied in exactly the same way to anyone buying a kilo of rice, whether rich or poor. If a person devotes his total income to the purchase of basic necessities, with a VAT of 18% he pays a tax of 18% on his total income. If, on the other hand, a person earning a very comfortable income spends only 10% of it on such purchases, the tax levied via VAT represents 1.8% of his income, leaving the rest of it free to be invested, and untaxed into the bargain!

• Massive privatisation of public companies, in other words the disinvolvement of government from competitive sectors of production. The forced privatisation of public companies at slashed prices directly benefits the Northern multinationals and a few individuals close to government, while the proceeds go straight towards repayment of the debt. In the case of Mali, of the 90 public companies in operation in 1985, 26 were liquidated and 28 were privatised: in 2001 only 36 remained. In the IMF's book, the government has no business in a sector where profits can be made. It should restrict itself to the more lofty functions (police, armed forces, justice) and withdraw from the other sectors (water, telecommunications, transportation, health, education, etc.). Thus government is voided of its political essence and at the same time loses control of the strategic elements for development. For the people, privatisations mean reduced access to services and increased unemployment.

SAPs: What Results?

With the hindsight of some twenty years, it can be seen that the macro-economic criteria favoured by the IMF and the World Bank have in no way helped improve the well-being of the poorest populations. On the contrary, greater impoverishment can be observed in all regions that have meekly applied structural adjustment, as well as increased inequality, not only within the various nations but between them. The narrow-sightedness of these two international institutions has resulted in complete failure in terms of human development.

This failure can in no way be imputed to an unfavourable economic situation or to lack of understanding, but rather to the strict application of neo-liberal policy. One can only wonder why measures so prejudicial to local populations are imposed with such constancy and diligence. The answer can be found in the words of Joseph Stiglitz, chief economist of the World Bank from 1997 to 1999, and 2001 Nobel laureate in economics: "*The IMF has ceased to serve the interests of the world economy in order to serve those of world finance. The liberalisation of the financial markets has not perhaps contributed to global economic stability, but it has certainly opened up immense new markets on Wall Street. (…) If one examines the IMF as if its objective was to serve the interests of the financial community, then one can make sense of actions that would otherwise appear contradictory and intellectually inconsistent*".[8]

Development aid, brandished by the wealthy countries as proof of goodwill, is quite unable to make good the losses incurred by debt: in 2003, for the 373 billion dollars that developing countries as a whole disbursed to service the debt,[9] they received only 69 billion dollars in public development aid.[10] In short, for every dollar entering these countries as aid, 5 dollars go out to pay back the debt. The World Bank's 2004 Global Development Finance report reveals moreover that the net transfer on debt—that is, the difference between new loans and the sums disbursed—is negative for the total of developing countries: on average, -90 billion dollars per year between 1998 and 2003. The debt is therefore a powerful mechanism for transfer of wealth from the populations of the South to their wealthy creditors. The United Nations Conference for Trade and Development (UNCTAD) stated in September 2004 that the African debt "amounts to a reverse transfer of resources from the world's poorest continent".[11]

Structural adjustment plans are perfectly capable of defending the interests of financial institutions and multinationals in the North. But for the people suffering the consequences, they are instruments of poverty and deprivation. The pill is a bitter one. As a complement to this study, see reference 12.

The Impact of Structural Adjustment Programmes on Malaria

The impact of SAPs on malaria can be described at various levels. The first, and the most direct, is the impact on determining health factors. The Third World's ailing healthcare systems are hard hit by the inroads made in social budgets to repay the debt: on an average, only 7% of Third World budgets go towards health, as against 21% for servicing the debt, and as much as 38% for Sub-Saharan Africa.[13] In material terms, this means the closing of healthcare structures, laying off of staff or wage reductions (to a level below the vital minimum), the deterioration of infrastructures, reduction of preventive care, gradual decay of health institutions at all levels (thus compromising effective healthcare policy-making by government), etc. Yet the responsibility of these destructive policies, imposed by the Bretton Woods institutions, has never been questioned: the acknowledgement of a failing healthcare system is always taken out of context and even used as an argument for further State disinvolvement. In this respect, the fact that the World Bank is today the institution that determines the shape of global healthcare, drawing on resources far superior to those of the WHO, is very revealing.

The commercialisation of health services is another angle of attack, driven either by the application of cost recovery policies, or simply by the mechanics of privatisation. In both instances, access to healthcare is endangered. One example among many: in 1995, by order of the IMF and the World Bank, Madagascar introduced PFU (user charging): "*a modest contribution by the patient towards the cost of the consultation and the medication. The average cost of a consultation in a healthcare centre (including medication for three days of treatment) is 400 to 500 ariary (28 to 36 euro cents). On July 20, 2002 Marc Ravalomanana, President of the Republic, ordered the suspension of PFU to prevent the political-military crisis in the first quarter of 2002 from overly affecting people's access to healthcare. According to a study made by the National Institute of Statistics and Cornell University (USA) in December 2002, this suspension increased visits to healthcare centres by 57% as compared to before the crisis*".[14] [AFP dispatch, 16 September 2003.] Unfortunately, the Ravalomanana régime, forced to bow to the logic of structural adjustment, decided to reinstate PFU in January 2004. Other studies tend to the same conclusions, and it is today an established fact that privatisation and cost recovery policies in the area of healthcare deprive a large section of the public of access to its services.

Excessive trade liberalisation as advocated by SAPs is also closely related to the resurgence of malaria. Take the famous "white elephants" - colossal, vastly expensive projects designed to connect developing countries with the world market by facilitating extraction and transportation of their raw materials. The great dams, already totally inappropriate to the people's needs, provided the anopheles that carry the disease with new ecological niches. In Sudan, the Gezira water management project effectively increased prevalence of the disease in the late 70s.[15] In addition, through the massive development of agricultural exports, SAPs endanger the principle of food security (the abandoning of basis food crops), and beyond that, cause drastic remodelling of the landscape and major modification of the ecosystem. At the end of the 1980s, the historian of medicine Randall Packard revealed that in Swaziland, irrigation practices—resulting from an economy directed toward sugar exports—had caused the reintroduction of malaria in areas where the disease was previously under control.[16] The use of pesticides is another aspect of sugar-exporting economies in developing countries: the drive to improve yield to increase export volume. Several studies have shown that excessive use of pesticides contributes to the formation of ecological niches where anopheles resistant to pesticides have permitted the resurgence of malaria.[17]

The impact of these various measures on malaria is considerable. In Vietnam, a WHO study published in a World Bank document showed that the number of deaths due to malaria tripled in the four years of the SAP, while in the same the healthcare system was seen to decline and the price of medicines soared.[18,19]

Beyond the harmful consequences of SAPs in the health sector, the close relationship between malaria, poverty and social justice must also be noted. Although it is not acknowledged, problems concerning housing, sanitary infrastructures and malnutrition among impoverished populations have a profound impact on their vulnerability to infection and on the pandemic in general.

What Solutions?

As we have seen, the official angle of approach to malaria is almost exclusively medicoscientific. Such an approach restricts the analysis of causes and establishes the framework for research on this single level, without taking into account the underlying politicoeconomic factors. The institutional consequence is the creation of vertical programmes, targeting only specific diseases, to the detriment of systemic programmes and primary healthcare. For example, the Global Health Fund was set up with the aim of raising funds to fight three major pandemics - an approach that fails to take in the social, political and global economic context. Under these conditions, and from the institutional standpoint, scientific research is seen as the only source of future solutions. But scientific research is itself a victim of the logic of structural adjustment, in the North as in the South: funds are constantly being cut back and research orientations are determined in terms of profitability. The field of tropical diseases, where no big profits can be made, is neglected in favour of research on cardio-vascular diseases and obesity, potentially very profitable areas.

Certain research projects are focussing on the use of genetically modified organisms (GMOs) to fight malaria. While we make no judgement on their effectiveness, which can only be assessed by scientists competent in this field, they should be regarded with the utmost caution. For several reasons. First, their innocuousness has yet to be proven, and therefore every precaution must be taken. In addition, they are a serious threat to biodiversity, and pose the problem of intellectual property.

Finally, and most importantly, a major reason why we are firmly opposed to the use of GMOs is that there are other effective, accessible, environmentally neutral and inexpensive means of fighting malaria (the distribution and reimpregnation of mosquito netting, access to curative care for timely treatment of suspected or confirmed cases, affordable medication, reduction of anopheles niches, etc.). The regrettable concentration on hyper-specialised scientific research and on vertical programmes as the only tools for counteracting the pandemic is part and parcel of a political and economic model. Given the state of proof so far (or rather the absence of same) guaranteeing that GMOs have no negative effects, we wish to express our categorical opposition to their use.

Conclusion

Why play the sorcerer's apprentice when simple solutions exist that have not been given a serious chance? Could the fact that they are neglected in favour of other solutions be related to the fact that they are out of phase with private financial interests and on the fringe of the dominant system, which consists of merchandising whole sections of the economy? Is this neo-liberal model in any way related to the failure to eradicate malaria – a goal that seemed globally feasible in the 1970s? By asking these questions in the light of the facts stated above, are the answers not patently obvious?

Legitimate and necessary as they may be, scientific solutions can only provide a partial response unless combined with a profound consideration of the socio-economic factors that underlie and shape the global malaria scene. The many human catastrophes caused by forcibly imposing wide-scale neo-liberal policies, of which the resurgence of malaria is a manifestation, should urge us to reexamine the economic world order. All aspects of today's macro-economic model must be questioned and radically changed if we want to influence the trend. We must give governments the means to set up and implement solid, multisector programmes to fight malaria, and above all, the means to improve living conditions for the greatest possible number. It requires a reversal of the current logic - a logic we must endeavour to unravel by every means possible.

Acknowledgements

Many thanks to Judith Harris for her precious help in translating this work.

References

1. http://www.rbm.who.int/cmc_upload/0/000/015/372/RBMInfosheet_1.htm http://www.rbm.who.int/cmc_upload/0/000/015/370/RBMInfosheet_3.htm (accessed March 14, 2005).
2. http://www.rbm.who.int/cmc_upload/0/000/015/370/RBMInfosheet_3.htm (accessed March 14, 2005).
3. http://www.rbm.who.int/cmc_upload/0/000/015/363/RBMInfosheet_10.htm (accessed March 14, 2005).
4. Abuja Declaration, Roll Back Malaria Summit, Abuja, Nigeria, 2000.
5. http://msf.fr/site/site.nsf/pages/2millions (accessed March 14, 2005).
6. Motchane JL. Si l'OMS voulait changer le monde. Apartheid Médical, Manière de voir, Le Monde Diplomatique 2004; 73:86-87.
7. Brundtland GH. Why invest in health? Third International Conference on Priorities in Health Care, Amsterdam, The Netherlands: 2000.
8. Stiglitz J. Globalization and its discontents. WW Norton and Company 2002.
9. World Bank, Global Development Finance Harnessing Cyclical Gains for Development. Washington DC: World Bank, 2004:650.
10. Final Official Development Assistance (ODA) data for 2003:5 (http://www.oecd.org/dataoecd/19/52/34352584.pdf).
11. UNCTAD Economic Development in Africa. Debt sustainability: Oasis or mirage? Geneva 2004.
12. Millet D, Toussaint E. Who owes who? 50 Questions about World Debt. London: Zedbooks, 2004:185.
13. Figure provided by Kofi Annan at the Okinawa G7 summit in 2000.
14. AFP dispatch, 16 September 2003.
15. Gruenbaum E. Struggling with the mosquito: Malaria policy and agricultural development in Sudan. Med Anthropo 1983; 7:51-62.
16. Packard RM. Agriculture development, migrant labour, and the resurgence of malaria in Swatziland. Soc Sci Med 1986; 22:861-867.
17. Chapin G, Wasserstorm R. Pesticide use and malaria resurgence in Central America and India. Soc Sci Med 1983; 17:273-290.
18. World Bank. Vietnam, Population, Health and Nutrition Review. Washington, DC: World Bank, 1993.
19. Chossudovsky M. The Globalisation of poverty. Impact of IMF and World Bank Reforms. Montréal: Ecosociété, 1997:248.

CHAPTER 3

Development of a Toolkit for Manipulating Malaria Vectors

Flaminia Catteruccia, Anthony E. Brown, Elisa Petris, Christina Scali and Andrea Crisanti*

Summary

This chapter will review the efforts made by several laboratories to generate in *Anopheles* mosquitoes a variety of molecular tools, similar to those available to the *Drosophila* community, to exploit the knowledge originating from the *A. gambiae* genome sequence. In particular, we will describe the experiments that have led to the development of gene transfer technologies for *Anopheles*, based on the use of efficient transposable elements to mediate germline integration of foreign genes and reliable selectable markers to detect positive transformants. This set of tools available for malaria research will be invaluable to improve our understanding of mosquito-*Plasmodium* interactions as well as to develop new vector-control measures. Functional studies to identify mosquito genes crucial for parasite development are being performed using RNA interference. At the same time, the post-integration behaviour of transposable elements is being investigated to assess their potential for insertional mutagenesis studies. It can be envisaged that in the near future transgenic technologies will assist in the implementation of malaria-control programs based on the eradication of vector species or their replacement with malaria-refractory populations.

A Historical Perspective to Mosquito Control:
A Case for Genetic Manipulation

Despite massive efforts to develop efficient malaria control programs, the number of malaria cases is on the rise, and is projected to double over the next 20 years if effective control measures are not introduced.[1] In the 1950s and early 1960s, the WHO launched a massive campaign for malaria eradication by using indoor spraying of DDT and large mass drug administration of chloroquine and pyrimethamine. In some cases these programs lacked epidemiological skills and knowledge, as well as administrative organization. As time progressed, it became evident that it was difficult to establish effective surveillance in the absence of a solid health infrastructure. Some of the factors that limited the success of these programs included the insurgence of DDT resistance in vector mosquitoes and the lack of logistic infrastructures. In the great majority of countries, eradication was not a realistic goal and there was a need to change from highly prescriptive, centralized control programs to flexible, cost-effective, and sustainable programs adapted to local conditions and responding to local needs.[2,3] This prompted research in novel control strategies, based on multi-disciplinary or indeed innovative approaches.

---

*Corresponding Author: Andrea Crisanti—Imperial College London, Department of Biological Sciences, Imperial College Road, London SW7 2AZ, U.K. Email: acrs@imperial.ac.uk

---

Genetically Modified Mosquitoes for Malaria Control, edited by Christophe Boëte.
©2006 Landes Bioscience.

The idea of controlling vector-borne diseases by a genetic means was proposed more than four decades ago[4,5] but has only recently become a concrete possibility. In the past few years, remarkable progress has been made in the development of efficient molecular and genetic tools available for *Anopheles* mosquitoes, the sole vectors of human malaria. This effort, prompted by the need to better understand the genetic basis of mosquito-pathogen compatibility, has led to the achievement of prominent milestones. Genetic manipulation of malaria vectors had been attempted for two decades, following the development of an efficient transformation system in *Drosophila melanogaster* based on the use of the *P* transposable element.[6] The massive research effort produced to identify the crucial components of a universally suitable transposition system culminated in the first germline transformation of *A. stephensi* mosquitoes,[7] followed by the development of transgenic *A. gambiae* lines.[8] The parallel completion of the *A. gambiae* genome sequence has then provided the missing link to exploit genetic manipulation to perform functional studies in mosquito vectors of human malaria and develop novel vector-control strategies.[9]

The Road to the Development of an *Anopheles* Transformation System

Transposable Elements

Mobile genetic elements called transposable elements or transposons have been used as tools to achieve genetic modification of many insect species. Insect transgenesis is a field that has impressively expanded in the last two decades thanks to the development of techniques to rapidly evaluate transposon mobility in nonhost insect species, significant advances in transposon delivery technique, and the discovery of reliable molecular markers essential for the identification of transformed individuals in insect species for which markers that complement eye pigment mutants have not been found.

The first germline transformation of an insect species was performed in 1982, when the *P*-element was shown to be able to integrate into the genome of *D. melanogaster*.[6] In the following years, efforts were made to export the *P*-element system to other insect species including mosquitoes. However it soon became apparent that for its activity, *P* requires cofactors only present in *Drosophila*. In fact, *P*-element activity decreases as a function of relatedness to *D. melanogaster* and no bona fide transposition activity has ever been observed outside the *Drosophilidae*.[10,11] There have been three reported attempts to introduce foreign genes into the mosquito genome by exploiting the *P* element. Miller and collaborators transformed *A. gambiae* with a *P*-based plasmid containing a gene coding for resistance to the neomycin analogue G-418.[12] They were able to maintain and study the integration for more than 50 generations. However transformed lines were obtained at very low frequencies (0.1% of injected embryos) and integration was mediated by illegitimate recombination rather than by the *P*-element. *Aedes aegypti*[13] and *Ae. triseriatus*[14] mosquitoes were also transformed using similar protocols but in both cases integration occurred at very low frequencies and was not *P*-mediated. Furthermore transgenic lines were lost in the second generation and analyses were not complete. Even if between 1985 and 1990 there were numerous attempts to use *P*-elements in nondrosophilid insects, they have remained unpublished presumably because transgenic insects were not obtained. The failure to use *P* to mediate integration of exogenous DNA into the mosquito germline prompted research on the identification of transposable elements with a broader host range. For this purpose the Interplasmid Transposition Assay (ITA) was developed to rapidly screen candidate transposons by assessing their mobility in the cellular environment of nonhost insect species.[15] This assay consists of the simultaneous injection of three plasmids into the target embryo. Excision of a transposon from a donor plasmid, and subsequent integration into a target plasmid mediated by the transposase gene expressed by a helper plasmid, can be detected using a combination of antibiotic resistance genes. ITA has allowed high-throughput screens, showing to be a good predictor for the ability of a transposon to mediate germline integration events. For example, a *Hermes* transposon from *Musca domestica* and a *mariner* element from *D. mauritiana* were shown to be capable of mobility in *Ae. aegypti* embryos[16,17] and subsequently mediated the first stable integration of exogenous DNA into the *Ae. aegypti* germline.[17,18] The *minos* transposable element from *D. hydei* was the first transposon shown to mediate germline transformation of a nondrosophilid insect, the Mediterranean medfly *Ceratitis capitata*.[19] Like all *Tc1*-like elements, *minos* inserts into many different sites of the genome and it requires only the presence of a TA dinucleotide at the target site, which is duplicated upon insertion. An Interplasmid Transposition Assay demonstrated that *minos* was capable of transposing in *A. stephensi* embryos, and *minos*-mediated integrations were obtained in *A. gambiae* cell lines.[20] These results were shortly followed by the first germline transformation of a human malaria vector, when *minos* was successfully shown to integrate into the genome of the Indian vector *A. stephensi*.[7] The following year, transformation of the principal vector of human malaria *A. gambiae* was achieved using the *piggyBac* transposon from *Trichoplusia ni*.[8] A *piggyBac* vector had first been used to transform the medfly *C. capitata*,[21] and since then it has been used to transform a variety of insect species spanning three orders. *PiggyBac* inserts at a TTAA target nucleotide sequence, which is duplicated upon insertion. The broad host-range of this transposon has allowed germline transformation of other mosquito species such as *Ae. aegypti* and *A. albimanus*.[22,23] In addition, mobility of the *piggyBac* transposon has also been demonstrated in embryos of *Ae. albopictus*, the vector of Dengue fever, and *Ae. triseriatus*, the vector of La Crosse encephalitis,[24] suggesting that germline transformation of these species will be achieved in the near future.

Selectable Markers

Initial attempts to generate transgenic mosquitoes were also hampered by the lack of suitable molecular markers of transformation. The first generation of markers for *Anopheles* transformation consisted of genes conferring resistance to antibiotics, such as the neomycin analogue G-418. However resistance to antibiotics varies considerably among individuals within the same wild type populations, and such markers have been shown to be unreliable and toxic to the cells. The germline transformation of the first nondrosophilid insect, the medfly *C. capitata*, was achieved only when a suitable visible selectable marker was identified. In *C. capitata* a null mutation in the *white eye* locus could be complemented by a cloned wild-type copy of this gene.[25] In 1998, the first transgenic *Ae. aegypti* mosquito lines were developed using as a selectable marker the *D. melanogaster cinnabar* gene[26] to rescue the white eye phenotype of an *Ae. aegypti* mutant strain (*kh*^w).[17,18] This white-eye (*w*) mutation is caused by a defect in the kynurenine hydroxilase enzyme that converts kyrunenine to 3-hydroxykynurenine.

As genes capable of rescuing visible mutant phenotypes in *Anopheles* have not yet been identified, a crucial step towards the development of transgenic technologies for the vectors of human malaria was the identification and manipulation of a series of fluorescent proteins derived from jellyfish or from corals. In a short time, this new generation of visible selectable markers has revolutionized the field of insect germline transformation. The first fluorescent protein to be used in germline transformation was the green fluorescent protein GFP from the jellyfish *Aequorea victoria*.[27] GFP is a protein of 238 amino acid residues, with a large absorbance peak at 395 nm and a smaller peak at 475 nm. Excitation at 395 nm yields an emission maximum at 508 nm. GFP was successfully used as a vital marker in *D. melanogaster*.[28] However, attempts to use GFP to transform *A. stephensi* mosquitoes failed, probably due to the instability of this protein (Catteruccia unpublished results). Since then a battery of red-shifted GFP mutants have been developed with improved qualities. The enhanced green fluorescent protein EGFP proved to be a reliable marker in the first germline transformation of *A. stephensi* mosquitoes,[7] and has since been used to transform *A. gambiae*[8] and *A. albimanus*[23] as well as *Culex quinquefasciatus*,[29] and *Ae. aegypti*.[30] Two stable mutants of GFP, the cyan fluorescing variant ECFP and the yellow fluorescent protein EYFP, have been used as transformation markers in *Drosophila* and can be distinguished using specific filter sets.[31] Currently, the use of EYFP is being validated in mosquito transformation after being successfully used in *A. gambiae* cell culture experiments (Catteruccia unpublished data). Importantly, the red fluorescent protein DsRed from the coral *Discosoma* has also been validated as a selectable marker for anopheline mosquitoes.[32] The availability of a second marker is crucial for studies based on multi component genetic systems, such as the UAS/GAL4 system and transposon remobilization experiments. The availability of flexible vector systems based on *Hermes*, *mariner* and *piggyBac* transposons, has also facilitated the transformation of a wide range of arthropod species.[31]

Functional Genomics in *Anopheles*

In this post-genomic era, with the genome sequences of most biologically important organisms complete, the research focus is on understanding the functions of individual genes on a genome-wide level. The number of known *Anopheles* genes is estimated to be in the region of 13,000–14,000, and a research priority is to understand the biochemical role of each of these genes and how their functions interact. In *D. melanogaster*, functional studies are usually performed randomly with chemical mutagens such as ethane methyl sulfonate[33] and through remobilisation of integrated transposons.[34,35] The genetic characterisation of nearly 60% of all the genes in *Drosophila* has been achieved using one or other of these approaches.[36,37] Although chemically-induced mutagenesis is not practical on a genome-wide scale due to the difficulty in identifying such mutations at the DNA level, transposon-mediated insertional mutagenesis has proven to be extremely useful. Insertional mutagenesis involves the mobilization of transposons into new chromosomal positions and the disruption of gene activity in the associated locus. Usually a marked, nonautonomous 'mutator' element is mobilized by a 'jumpstarter' element providing, in trans, the transposase activity. Enhancer elements can also be isolated in such screens, when integration of a reporter gene allows the identification of a tissue-specific pattern of expression. Such applications require high rates of remobilization in the presence of active transposase.

P element-mediated mutagenesis has been used in *Drosophila* to mutate ~25% of genes essential for adult viability.[38,39] This has greatly increased our understanding of fruitfly behaviour and biology, leading to the identification of many genes involved in development, immunity, tissue modelling and embryogenesis.[34,40,41] However, *P* elements do not distribute randomly within or between genes but favour certain "hotspots", frequently integrating near or within gene promoters.[41] This characteristic of *P elements* makes saturation mutagenesis difficult and results in undesirable hypomorphic mutations. The use of other transposable elements has helped to overcome this limitation. Recently, *piggyBac* was shown to be as efficient as the *P*-element in transposon-mediated mutagenesis. Using *piggyBac* as a second mutagen, whose insertion profile is significantly more random than that of *P*, especially regarding "hot" and "cold" areas of the genome, inserts in essential genes that were previously recalcitrant have been generated.[37,42,43] Novel autosomal insertions have been recovered with an average jumping rate of 80%.[43] Insertions were located in both coding and noncoding regions of already characterized genes and in uncharacterized and non *P*-element-targeted genes. A total of nine novel transposon-mediated lethal insertions were identified. Furthermore, an enhancer-detection system was also able to detect head-specific, thorax-specific, abdominal-specific and leg-specific enhancers. The *Hobo* transposon has similarly been used in transformation and enhancer trapping, and shown to have different insertion specificity from that of *P*.[44] Addition of more transposable elements to this kind of study is likely to push these screens towards saturation.

The introduction of *P*-mediated mutagenesis in *Drosophila* has served as a paradigm for developing similar methodologies in other organisms. Studying the post-integration behaviour of transposons in insect species is important to predict whether they would represent useful tools in insertional mutagenesis studies. In *Anopheles*, large-scale insertional mutagenesis studies are constrained by practical and technical issues that limit their use. A mosquito's obligation for blood feeding to propagate the next generation, its aquatic larval developmental stage and the lack of single pair mating all combine to make large-scale remobilisation screens challenging in time and space. However, remobilization of integrated transposons could serve an important role in smaller-scale studies. To date, in mosquitoes, data exists only on the remobilization potential of the *MosI* (*mariner*), *Hermes* and *piggyBac* vectors in *Ae. aegypti*. *MosI*-based gene vectors were found to have a very low remobilization potential in both the soma and germ-line of *Ae. aegypti*.[45] Few somatic transpositions were detected and only a single germ-line transposition event was identified in 14,000 progeny (0.01%), suggesting *MosI* may not be a good candidate for transposon-mediated insertional mutagenesis. Furthermore, of the transposition events detected, approximately 25% (4/17) had occurred into a copy of the transposon itself. *Hermes* transposition has also been investigated in *Ae. aegypti* using a marked autonomous transposon. *Hermes* was found to transpose in the somatic cells of *Ae. aegypti* using a typical cut and paste mechanism characteristic of other class II transposable elements; however, remobilization in the germ-line was not detected despite several attempts.[46] Although not well documented, *piggyBac* appears to be inefficiently mobilized in *Ae. aegypti*.[46] *Minos* remobilizations have also been investigated. In *D. melanogaster*, transposition of an X-linked *minos* transposon into new autosomal sites occurred in 1–12% of males, demonstrating its potential for use in transposon-mediated insertional mutagenesis.[47] Donor sites after excision of *minos* were repaired 3 fold more frequently with gap repair (homologous recombination) than ligation repair (nonhomologous recombination). Direct injection of transposase mRNA was found to greatly increase the efficiency of transposition (31.8%).[48] The transposase-mediated mobilization of a *minos* transposon in the presence of a chromosomal source of transposase is currently being investigated in the soma and germ-line of *A. stephensi* (Scali and Catteruccia, unpublished results).

Recently, a method to facilitate gene targeting by homologous recombination using a yeast recombinase was developed in *Drosophila*[49] and led to the targeting of more than 20 different loci,[49-52] most of which had not been previously identified in genetic screens.[51] Although its heterologous nature makes it readily transferable to *Anopheles* mosquitoes, such a targeting system is based on an extremely complex tripartite molecular mechanism that occurs at relatively low frequencies and takes about six months per target gene, making it too laborious for large-scale screens of gene function.

With the publication of the *A. gambiae* genome sequence, a more rapid link between sequence data and biological function can be established. Reverse genetic analysis, that is the disruption of the activity of a molecularly characterised gene and examining the phenotype of the resulting mutant, provides the opportunity to do this. It is now possible to perform targeted functional studies by using RNA interference (RNAi) technology, in a highly-sequence-specific, inexpensive and technologically feasible manner. A significant advantage of this technology is that it perfectly complements other functional genomic platforms. The use of DNA microarrays for example, allows the entire set of ~13,000–14,000 *A. gambiae* genes to be represented in a single slide and the transcriptome to be analysed for changes in gene expression in the mosquito in response to parasite development. Those genes that are upregulated upon parasite infection could be the focus of follow-up functional screens and be specifically targeted (rather than the fortuitous knockout mediated by insertional mutagenesis).

RNA Interference: A Universal Approach to Functional Genomics

RNA interference was first discovered in the nematode *Caenorhabditis elegans*[53] but has since been show to be active in organisms as diverse as plants,[54] fungi,[55] mammals,[56] and insects.[57] As such it has been universally adopted as the method of choice to perform functional genomic studies in organisms that, like *Anopheles*, have previously proven intractable to targeted gene manipulation and whose genomes remain largely uncharacterised.

Silencing is first initiated when long dsRNA trigger molecules, introduced into a cell by microinjection, electroporation or transfection, are processed into 21-23 nucleotide small interfering RNAs (siRNAs)[58-61] by the RNase III enzyme Dicer (Fig. 1, i).[62,63] These siRNAs are proposed to first assemble into a 360-kDa ribonucleotide protein complex (RNP) that transfers the siRNA to the RNA-induced silencing complex (RISC).[64] Upon activation with ATP,

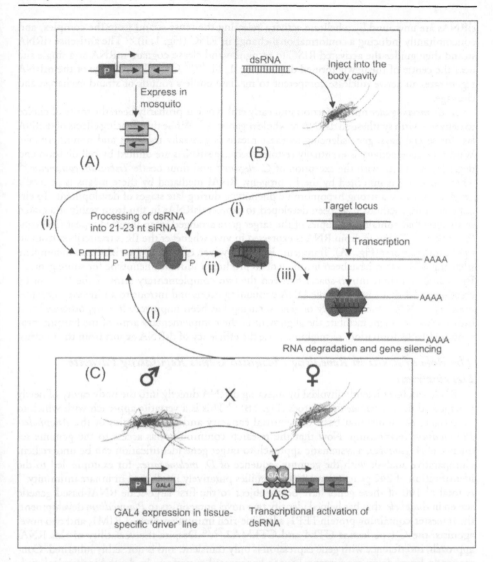

Figure 1. A versatile RNAi toolbox for *Anopheles* mosquitoes. Double-stranded RNA (dsRNA) homologous to the target gene is delivered by either of two methods: A) by the tissue-specific expression of an IR in vivo (P, tissue specific promoter), or B) by the injection of in vitro-transcribed dsRNA into individual mosquitoes. Transgenic RNAi can be further enhanced by utilising the *GAL4-UAS* system, directing dsRNA expression in specific tissues (C). In this system, one mosquito line (*green*) is developed that expresses the yeast transcriptional activator GAL4 under the control of a tissue-specific *Anopheles* promoter (P). A second transgenic line (*red*) is developed by inserting the upstream activating sequence (UAS), to which GAL4 binds, next to a target gene-specific IR. Crossing these two transgenic lines results in F_1 progeny that specifically express dsRNA in tissues where GAL4 is expressed. i) DsRNA delivered by each means is cleaved into small interfering RNAs (siRNAs) by Dicer (*red*). The siRNA/Dicer complex is then recruited by the ribonucleotide protein complex (RNP) (*blue*) and transferred to the RNA-induced silencing complex (RISC) (*purple*). ii) RISC becomes activated by unwinding of the siRNA duplex by an RNA helicase. iii) The sense strand dissociates from RISC and is replaced by the target mRNA complementary to the antisense siRNA strand. A nuclease activity then cleaves the mRNA, leading to post-transcriptional silencing of the target locus. A color version of this figure is available online at http://www.Eurekah.com.

siRNAs are unwound by a helicase activity, releasing the sense strand from the complex, and concomitantly inducing a conformational change in RISC (Fig. 1, ii).[65] The antisense siRNA strand then guides the activated RISC to recognise and cleave cognate mRNA at a single site near the centre of the newly formed duplex (Fig. 1, iii).[66,67] Crucially, cleavage of the mRNA regenerates an active nuclease competent to mediate endless rounds of strand exchange and cleavage.

In *D. melanogaster* microinjection into early embryos has primarily been the route of choice to deliver in vitro synthesised dsRNA to whole organisms.[57] While the silencing effects of dsRNA last for several days, gene silencing by these means is generally transient and noninheritable. Wild-type gene activity is eventually restored because siRNAs are diluted by cell division and degradation. Thus, with the exception of *C. elegans*[53] and flour beetle *Tribolium castaneum*,[68] where silencing is inherited by the F_1 progeny, RNAi mediated by these means is generally insufficient to see a strong hypomorphic phenotype during late stages of development. To circumvent this, methods have been developed to express dsRNAs in situ from stably integrated transgenes that contain two copies of the target gene arranged in an inverted repeat (IR) configuration.[69,70] Thus, hairpin RNA is expressed in vivo whenever the IR is transcribed from an upstream promoter (Fig. 1A). Transgenic RNAi, however, often shows variable and incomplete silencing effects that have been insufficient to produce a mutant phenotype for some genes.[71] By including a short intron-spacer between the two complementary arms of the IR,[72] or by creating an IR that fuses genomic DNA containing exons and introns to an inverted complementary cDNA,[71] the efficiency of gene silencing has been improved. It is hypothesised that intron excision might facilitate the alignment of the complementary arms of the hairpin, promoting dsRNA formation, or might increase the efficiency of dsRNA export from the nucleus.

The Role of RNAi in Revealing Mosquito Genes Regulating Parasite Development

RNAi can be efficiently invoked by injecting dsRNA directly into the body cavity of newly emerged adult *A. gambiae* mosquitoes (Fig. 1B).[73] This is a versatile approach with which to investigate the molecular basis of vectorial capacity and the dynamics of the *Anopheles-Plasmodium* relationship. Now that the research community has access to the genome sequence of *A. gambiae*, a systematic approach to target gene identification can be undertaken. Comparative analysis with the genome sequence of *D. melanogaster*, for example, led to the identification of 242 genes from 18 gene families putatively implicated in innate immunity.[74] A total of 100 of these genes have been subject to the first large-scale RNAi-based genetic screen in *Anopheles* that has helped identify two novel antagonists to *Plasmodium* development, the thioester-containing protein TEP1, a leucine-rich immune protein, LRIM1, and two novel agonists, the C-type lectins CTL4 and CTLMA2.[75,76] Despite the versatility of this RNAi approach, interference with gene expression is only transient, and is not stably inherited. Consequently, knockdown experiments have to be repeated to perform in-depth biochemical studies. By expressing dsRNA in situ in mosquitoes from stably integrated transgenes with dyad symmetry these limitations are overcome and an unlimited supply of uniformly mutant transgenic organisms is readily available for biochemical studies.

In validating heritable RNAi for *Anopheles*,[77] Brown et al generated transgenic *A. stephensi* mosquitoes stably expressing an RNAi transgene designed to produce intron-spliced dsRNA targeting the green fluorescent protein *EGFP* gene. Crossing these RNAi effector lines with an *EGFP*-expressing target line resulted in dosage-dependent *EGFP* silencing. Significantly this study also revealed that like *Drosophila*,[71,78] transgene-mediated gene silencing in *Anopheles* is strictly confined to those cells in which the target gene and dsRNA gene are coexpressed. Thus it seems that, while, in common with transgenic RNAi in *C. elegans*,[79] dsRNAs delivered from the extracellular environment can readily cross the cell membrane and invoke RNAi, when produced intracellularly from stably integrated transgenes, they cannot spread to neighbouring cells to perpetuate the silencing signal.

The cell autonomous nature of transgenic RNAi significantly extends its experimental possibilities. By placing dsRNA expression under the control of a tissue-specific promoter it should be possible to tailor it to coincide spatially and temporally with the development of the *Plasmodium* parasite in the mosquito. In this way gene expression in other tissues would not be affected, avoiding any additional pleiotropic effects this may bring. Expression of the RNAi transgene could be controlled by a bipartite system such as the *GAL4-UAS* system (Fig. 1C), or an inducible gene expression system such as 'Tet-On' and 'Tet-Off' systems recently developed for anopheline mosquitoes.[80] Thus, with analogy to the collection of enhancer trap lines that have been developed in *Drosophila*,[81] a library of driver lines could be established, each directing the expression of a transcriptional activator in a precisely defined manner, which could then be crossed to dsRNA responder stocks, permitting one to dissect multiple roles of the same gene in different tissues and at different developmental stages. One area of research that should benefit from this approach is the identification of specific molecules on the surface of mosquito tissues that function as receptors for parasite development.

Future Directions for Functional Genomics in Anopheles Mosquitoes

The completion of the *A. gambiae* genome project allows the possibility to perform genome-wide RNAi screens. Shortly after the *C. elegans* genome sequence became available, oligonucleotide primers were designed for generating dsRNAs against every protein-coding gene for use in genome-wide RNAi screens.[82-84] A similar strategy has recently been employed to undertake genome-wide screens in *Drosophila* cell culture.[85] With the development of a whole transcriptome microarray for *Anopheles*, a collection of templates from which to synthesise dsRNAs will soon be available. It remains to be seen, however, whether current protocols based upon microinjection will be appropriate for such a large undertaking.

Transgenic Technologies for Malaria Control

The development of new genetic and molecular tools for *Anopheles* mosquitoes has provided new strength to malaria eradication programs based on vector control. Transgenic technologies could be used to replace natural disease-transmitting populations with genetically modified mosquitoes (GMM) refractory to malaria parasites, or to suppress or eradicate field populations by introducing large numbers of GMM sterile males in Sterile Insect Technique (SIT) programs. A series of effector mechanisms have been proposed to block malaria transmission in the mosquito vector. Using genetic transformation, modifications have already been introduced which demonstrate a capacity to interfere with malaria development. The bee venom phospholipase A2 (PLA2) has been shown to strongly inhibit oocyst formation by interfering with ookinete invasion of the midgut epithelium of transgenic *A. stephensi* mosquitoes, possibly by modifying the properties of the midgut epithelial membranes that are invaded by the parasite.[86] An artificial peptide SM1 (for salivary gland-and midgut-binding peptide 1) strongly inhibited the crossing of the midgut epithelium by *Plasmodium* parasites.[87] Concerns have however been raised over the safety, ethical and efficacy issues related to the release of such effector molecules in the field.[88,89]

The alternative strategy of developing SIT for malaria control is gaining some support in the mosquito community. SIT is a species-specific method of insect control that depends on the mass rearing, sterilization and release of large numbers of male insects into the general population.[4,5] As the mating of sterile males with wild type virgin females would produce no progeny, if large numbers of sterile males are released over a sufficient period of time, and the percentage of multiple matings is low, the local eradication of the pest population will ensue.

The paradigm of area-wide SIT programs has been the successful elimination of the New World screw worm, *Cochliomyia hominivorax*, from the southern states of the USA, Mexico and all of Central America.[90] This area is now protected from reinvasion from South America by the release of relatively small numbers of sterile flies across a narrow barrier in Panama. Mosquito releases have also been performed for numerous purposes related to SIT. Many studies were targeted at answering a specific question and did not anticipate immediate population suppression. Several suppression studies have also been attempted, but the programs were not of sufficient scale to be effective in nonisolated areas. Regardless of species, various prominent technical causes contributed to failure of mosquito releases:

1. Production below desired levels due to absence of sexing strains or delays in production;
2. Loss of male fitness owing to sterilization technique;
3. Immigration of mated females into release areas.

It must be stressed that different vector species need to be targeted in order to achieve suppression of the malaria parasite, rendering the application of SIT in malaria-control programs more complicated than the eradication of the screwworm. Moreover, multiple matings of female mosquitoes have been reported in the field, which could impair the efficacy of SIT programs.[91] However, SIT could be successfully implemented in areas where there exist a simple vector-parasite relationship and where the immigration of mated females or other vector species is not likely to occur.

Transgenic technologies have been proposed to help the generation of effective genetic sexing strains and to induce sterility in males without the need for irradiation. Current methods of separation of the two sexes are based on inefficient procedures such as the elimination of females based on the size of the pupae and pseudo-linkage of sex chromosomes to insecticide resistance and pupa colour alleles. Genetic manipulation of the mosquito genome could allow the development of reliable sexing strains and induce sterility in males without affecting mating competitiveness, making SIT programs far more powerful. It is feasible that targeting genes specific to spermatogenesis (the generation of mature sperm cells) or the sex-determination pathway could generate sterile male mosquitoes, which have a higher level of competitiveness than irradiated males. It is safe to predict that in the next few years SIT will benefit from a battery of novel molecular and genetic tools for the development of efficient genetic sexing strains and competitive sterile male mosquito populations. Regardless of the control strategy implemented, questions concerning horizontal transfer, toxicity against nontarget species and, in the case of SIT programs, effects of mosquito eradication on the food chain, will need to be scrupulously addressed, both in the laboratory and in caged trials, before a release of transgenic mosquitoes is performed.[92]

References

1. Breman JG, Egan A, Keusch GT. The intolerable burden of malaria: A new look at the numbers. Am J Trop Med Hyg 2001; 64(1-2 Suppl):iv-vii.
2. Global malaria control. WHO Malaria Unit. Bull World Health Organ 1993; 71(3-4):281-284.
3. Implementation of the global malaria control strategy. Report of a WHO study group on the implementation of the global plan of action for malaria control 1993-2000. World Health Organ Tech Rep Ser 1993; 839:1-57.
4. Knipling EF. Sterile-male method of population control. Science 1959; 130:902-904.
5. Knipling EF, Laven H, Craig GB et al. Genetic control of insects of public health importance. Bull World Health Organ 1968; 38(3):421-438.
6. Rubin GM, Spradling AC. Genetic transformation of Drosophila with transposable element vectors. Science 1982; 218(4570):348-353.
7. Catteruccia F, Nolan T, Loukeris TG et al. Stable germline transformation of the malaria mosquito Anopheles stephensi. Nature 2000; 405(6789):959-962.
8. Grossman GL, Rafferty CS, Clayton JR et al. Germline transformation of the malaria vector, Anopheles gambiae, with the piggyBac transposable element. Insect Mol Biol 2001; 10(6):597-604.
9. Holt RA, Subramanian GM, Halpern A et al. The genome sequence of the malaria mosquito Anopheles gambiae. Science 2002; 298(5591):129-149.
10. Handler AM, O'Brochta DA. Prospects for gene transformation in insects. Annu Rev Entomol 1991; 36:159-183.
11. O'Brochta DA, Gomez SP, Handler AM. P element excision in Drosophila melanogaster and related drosophilids. Mol Gen Genet 1991; 225(3):387-394.
12. Miller LH, Sakai RK, Romans P et al. Stable integration and expression of a bacterial gene in the mosquito Anopheles gambiae. Science 1987; 237(4816):779-781.

13. Morris AC, Eggleston P, Crampton JM. Genetic transformation of the mosquito Aedes aegypti by micro-injection of DNA. Med Vet Entomol 1989; 3(1):1-7.
14. McGrane V, Carlson JO, Miller BR et al. Microinjection of DNA into Aedes triseriatus ova and detection of integration. Am J Trop Med Hyg 1988; 39(5):502-510.
15. Handler AM, Gomez SP, O'Brochta DA. A functional analysis of the P-element gene-transfer vector in insects. Arch Insect Biochem Physiol 1993; 22(3-4):373-384.
16. Sarkar A, Yardley K, Atkinson PW et al. Transposition of the Hermes element in embryos of the vector mosquito, Aedes aegypti. Insect Biochem Mol Biol 1997; 27(5):359-363.
17. Coates CJ, Jasinskiene N, Miyashiro L et al. Mariner transposition and transformation of the yellow fever mosquito, Aedes aegypti. Proc Natl Acad Sci USA 1998; 95(7):3748-3751.
18. Jasinskiene N, Coates CJ, Benedict MQ et al. Stable transformation of the yellow fever mosquito, Aedes aegypti, with the Hermes element from the housefly. Proc Natl Acad Sci USA 1998; 95(7):3743-3747.
19. Loukeris TG, Livadaras I, Arca B et al. Gene transfer into the medfly, Ceratitis capitata, with a Drosophila hydei transposable element. Science 1995; 270(5244):2002-2005.
20. Catteruccia F, Nolan T, Blass C et al. Toward Anopheles transformation: Minos element activity in anopheline cells and embryos. Proc Natl Acad Sci USA 2000; 97(5):2157-2162.
21. Handler AM, McCombs SD, Fraser MJ et al. The lepidopteran transposon vector, piggyBac, mediates germ-line transformation in the Mediterranean fruit fly. Proc Natl Acad Sci USA 1998; 95(13):7520-7525.
22. Kokoza V, Ahmed A, Wimmer EA et al. Efficient transformation of the yellow fever mosquito Aedes aegypti using the piggyBac transposable element vector pBac[3xP3-EGFP afm]. Insect Biochem Mol Biol 2001; 31(12):1137-1143.
23. Perera OP, Harrell IR, Handler AM. Germ-line transformation of the South American malaria vector, Anopheles albimanus, with a piggyBac/EGFP transposon vector is routine and highly efficient. Insect Mol Biol 2002; 11(4):291-297.
24. Lobo N, Li X, Hua-Van A et al. Mobility of the piggyBac transposon in embryos of the vectors of Dengue fever (Aedes albopictus) and La Crosse encephalitis (Ae. triseriatus). Mol Genet Genomics 2001; 265(1):66-71.
25. Zwiebel LJ, Saccone G, Zacharopoulou A et al. The white gene of Ceratitis capitata: A phenotypic marker for germline transformation. Science 1995; 270(5244):2005-2008.
26. Paton DR, Sullivan DT. Mutagenesis at the cinnabar locus in Drosophila melanogaster. Biochem Genet 1978; 16(9-10):855-865.
27. Chalfie M, Tu Y, Euskirchen G et al. Green fluorescent protein as a marker for gene expression. Science 1994; 263(5148):802-805.
28. Yeh E, Gustafson K, Boulianne GL. Green fluorescent protein as a vital marker and reporter of gene expression in Drosophila. Proc Natl Acad Sci USA 1995; 92(15):7036-7040.
29. Allen ML, O'Brochta DA, Atkinson PW et al. Stable, germ-line transformation of Culex quinquefasciatus (Diptera: Culicidae). J Med Entomol 2001; 38(5):701-710.
30. Pinkerton AC, Michel K, O'Brochta DA et al. Green fluorescent protein as a genetic marker in transgenic Aedes aegypti. Insect Mol Biol 2000; 9(1):1-10.
31. Horn C, Wimmer EA. A versatile vector set for animal transgenesis. Dev Genes Evol 2000; 210(12):630-637.
32. Nolan T, Bower TM, Brown AE et al. PiggyBac-mediated germline transformation of the malaria mosquito Anopheles stephensi using the red fluorescent protein dsRED as a selectable marker. J Biol Chem 2002; 277(11):8759-8762, (Epub 2002 Jan 8722).
33. Ashburner M. Drosophila: A laboratory handbook. Cold Spring Harbour Lab, 1989.
34. Cooley L, Kelley R, Spradling A. Insertional mutagenesis of the Drosophila genome with single P elements. Science 1988; 239(4844):1121-1128.
35. Robertson HM, Preston CR, Phillis RW et al. A stable genomic source of P element transposase in Drosophila melanogaster. Genetics 1988; 118(3):461-470.
36. Rubin GM, Lewis EB. A brief history of Drosophila's contributions to genome research. Science 2000; 287(5461):2216-2218.
37. Thibault ST, Singer MA, Miyazaki WY et al. A complementary transposon tool kit for Drosophila melanogaster using P and piggyBac. Nat Genet 2004; 36(3):283-287, (Epub 2004 Feb 2022).
38. Spradling AC, Stern D, Beaton A et al. The berkeley Drosophila genome project gene disruption project: Single P-element insertions mutating 25% of vital Drosophila genes. Genetics 1999; 153(1):135-177.
39. Peter A, Schottler P, Werner M et al. Mapping and identification of essential gene functions on the X chromosome of Drosophila. EMBO Rep 2002; 3(1):34-38, (Epub 2001 Dec 2019).

40. Wilson C, Pearson RK, Bellen HJ et al. P-element-mediated enhancer detection: An efficient method for isolating and characterizing developmentally regulated genes in Drosophila. Genes Dev 1989; 3(9):1301-1313.
41. Spradling AC, Stern DM, Kiss I et al. Gene disruptions using P transposable elements: An integral component of the Drosophila genome project. Proc Natl Acad Sci USA 1995; 92(24):10824-10830.
42. Hacker U, Nystedt S, Barmchi MP et al. PiggyBac-based insertional mutagenesis in the presence of stably integrated P elements in Drosophila. Proc Natl Acad Sci USA 2003; 100(13):7720-7725, (Epub 2003 Jun 7711).
43. Horn C, Offen N, Nystedt S et al. PiggyBac-based insertional mutagenesis and enhancer detection as a tool for functional insect genomics. Genetics 2003; 163(2):647-661.
44. Smith D, Wohlgemuth J, Calvi BR et al. Hobo enhancer trapping mutagenesis in Drosophila reveals an insertion specificity different from P elements. Genetics 1993; 135(4):1063-1076.
45. Wilson R, Orsetti J, Klocko AD et al. Post-integration behavior of a Mos1 mariner gene vector in Aedes aegypti. Insect Biochem Mol Biol 2003; 33(9):853-863.
46. O'Brochta DA, Sethuraman N, Wilson R et al. Gene vector and transposable element behavior in mosquitoes. J Exp Biol 2003; 206(Pt 21):3823-3834.
47. Arca B, Zabalou S, Loukeris TG et al. Mobilization of a Minos transposon in Drosophila melanogaster chromosomes and chromatid repair by heteroduplex formation. Genetics 1997; 145(2):267-279.
48. Kapetanaki MG, Loukeris TG, Livadaras I et al. High frequencies of Minos transposon mobilization are obtained in insects by using in vitro synthesized mRNA as a source of transposase. Nucleic Acids Res 2002; 30(15):3333-3340.
49. Rong YS, Golic KG. Gene targeting by homologous recombination in Drosophila. Science 2000; 288(5473):2013-2018.
50. Rong YS, Golic KG. A targeted gene knockout in Drosophila. Genetics 2001; 157(3):1307-1312.
51. Rong YS. Gene targeting by homologous recombination: A powerful addition to the genetic arsenal for Drosophila geneticists. Biochem Biophys Res Commun 2002; 297(1):1-5.
52. Seum C, Pauli D, Delattre M et al. Isolation of Su(var)3-7 mutations by homologous recombination in Drosophila melanogaster. Genetics 2002; 161(3):1125-1136.
53. Fire A, Xu S, Montgomery MK et al. Potent and specific genetic interference by double-stranded RNA in Caenorhabditis elegans. Nature 1998; 391(6669):806-811.
54. Akashi H, Miyagishi M, Kurata H et al. A simple and rapid system for the quantitation of RNA interference in plant cultured cells. Nucleic Acids Res Suppl 2001; (1):235-236.
55. Cogoni C, Macino G. Isolation of quelling-defective (qde) mutants impaired in posttranscriptional transgene-induced gene silencing in Neurospora crassa. Proc Natl Acad Sci USA 1997; 94(19):10233-10238.
56. Wianny F, Zernicka-Goetz M. Specific interference with gene function by double-stranded RNA in early mouse development. Nat Cell Biol 2000; 2(2):70-75.
57. Kennerdell JR, Carthew RW. Use of dsRNA-mediated genetic interference to demonstrate that frizzled and frizzled 2 act in the wingless pathway. Cell 1998; 95(7):1017-1026.
58. Hamilton AJ, Baulcombe DC. A species of small antisense RNA in posttranscriptional gene silencing in plants. Science 1999; 286(5441):950-952.
59. Tuschl T, Zamore PD, Lehmann R et al. Targeted mRNA degradation by double-stranded RNA in vitro. Genes Dev 1999; 13(24):3191-3197.
60. Zamore PD, Tuschl T, Sharp PA et al. RNAi: Double-stranded RNA directs the ATP-dependent cleavage of mRNA at 21 to 23 nucleotide intervals. Cell 2000; 101(1):25-33.
61. Elbashir SM, Lendeckel W, Tuschl T. RNA interference is mediated by 21- and 22-nucleotide RNAs. Genes Dev 2001; 15(2):188-200.
62. Hammond SM, Bernstein E, Beach D et al. An RNA-directed nuclease mediates post-transcriptional gene silencing in Drosophila cells. Nature 2000; 404(6775):293-296.
63. Bernstein E, Caudy AA, Hammond SM et al. Role for a bidentate ribonuclease in the initiation step of RNA interference. Nature 2001; 409(6818):363-366.
64. Hammond SM, Boettcher S, Caudy AA et al. Argonaute2, a link between genetic and biochemical analyses of RNAi. Science 2001; 293(5532):1146-1150.
65. Schwarz DS, Hutvagner G, Haley B et al. Evidence that siRNAs function as guides, not primers, in the Drosophila and human RNAi pathways. Mol Cell 2002; 10(3):537-548.
66. Caplen NJ, Parrish S, Imani F et al. Specific inhibition of gene expression by small double-stranded RNAs in invertebrate and vertebrate systems. Proc Natl Acad Sci USA 2001; 98(17):9742-9747.
67. Elbashir SM, Martinez J, Patkaniowska A et al. Functional anatomy of siRNAs for mediating efficient RNAi in Drosophila melanogaster embryo lysate. EMBO J 2001; 20(23):6877-6888.

68. Bucher G, Scholten J, Klingler M. Parental RNAi in Tribolium (Coleoptera). Curr Biol 2002; 12(3):R85-86.
69. Fortier E, Belote JM. Temperaturedependent gene silencing by an expressed inverted repeat in Drosophila. Genesis 2000; 26(4):240-244.
70. Kennerdell JR, Carthew RW. Heritable gene silencing in Drosophila using double-stranded RNA. Nat Biotechnol 2000; 18(8):896-898.
71. Kalidas S, Smith DP. Novel genomic cDNA hybrids produce effective RNA interference in adult Drosophila. Neuron 2002; 33(2):177-184.
72. Reichhart JM, Ligoxygakis P, Naitza S et al. Splice-activated UAS hairpin vector gives complete RNAi knockout of single or double target transcripts in Drosophila melanogaster. Genesis 2002; 34(1-2):160-164.
73. Blandin S, Moita LF, Kocher T et al. Reverse genetics in the mosquito Anopheles gambiae: Targeted disruption of the Defensin gene. EMBO Rep 2002; 3(9):852-856.
74. Christophides GK, Zdobnov E, Barillas-Mury C et al. Immunity-related genes and gene families in Anopheles gambiae. Science 2002; 298(5591):159-165.
75. Osta MA, Christophides GK, Kafatos FC. Effects of mosquito genes on Plasmodium development. Science 2004; 303(5666):2030-2032.
76. Blandin S, Shiao SH, Moita LF et al. Complement-like protein TEP1 is a determinant of vectorial capacity in the malaria vector Anopheles gambiae. Cell 2004; 116(5):661-670.
77. Brown AE, Bugeon L, Crisanti A et al. Stable and heritable gene silencing in the malaria vector Anopheles stephensi. Nucleic Acids Res 2003; 31(15):e85.
78. Billuart P, Winter CG, Maresh A et al. Regulating axon branch stability: The role of p190 RhoGAP in repressing a retraction signaling pathway. Cell 2001; 107(2):195-207.
79. Tijsterman M, May RC, Simmer F et al. Genes required for systemic RNA interference in Caenorhabditis elegans. Curr Biol 2004; 14(2):111-116.
80. Lycett GJ, Kafatos FC, Loukeris TG. Conditional expression in the malaria mosquito Anopheles stephensi with Tet-On and Tet-Off systems. Genetics 2004; 167(4):1781-1790.
81. O'Kane CJ, Gehring WJ. Detection in situ of genomic regulatory elements in Drosophila. Proc Natl Acad Sci USA 1987; 84(24):9123-9127.
82. Fraser AG, Kamath RS, Zipperlen P et al. Functional genomic analysis of C. elegans chromosome I by systematic RNA interference. Nature 2000; 408(6810):325-330.
83. Gonczy P, Echeverri C, Oegema K et al. Functional genomic analysis of cell division in C. elegans using RNAi of genes on chromosome III. Nature 2000; 408(6810):331-336.
84. Piano F, Schetter AJ, Mangone M et al. RNAi analysis of genes expressed in the ovary of Caenorhabditis elegans. Curr Biol 2000; 10(24):1619-1622.
85. Boutros M, Kiger AA, Armknecht S et al. Genome-wide RNAi analysis of growth and viability in Drosophila cells. Science 2004; 303(5659):832-835.
86. Moreira LA, Ito J, Ghosh A et al. Bee venom phospholipase inhibits malaria parasite development in transgenic mosquitoes. J Biol Chem 2002; 277(43):40839-40843.
87. Ito J, Ghosh A, Moreira LA et al. Transgenic anopheline mosquitoes impaired in transmission of a malaria parasite. Nature 2002; 417(6887):452-455.
88. Boëte C, Koella JC. Evolutionary ideas about genetically manipulated mosquitoes and malaria control. Trends Parasitol 2003; 19(1):32-38.
89. Scott TW et al. The ecology of genetically modified mosquitoes. Science 2002; 298(5591):117-119.
90. Wyss JH. Screwworm eradication in the Americas. Ann NY Acad Sci 2000; 916:186-193.
91. Tripet F et al. Frequency of multiple insemination in field-collected Anopheles gambiae females revealed by DNA analysis of transferred sperm. Am J Trop Med Hyg 2003; 68(1):1-5.
92. Coleman PG, Alphey L. Genetic control of vector populations: An imminent prospect. Trop Med Int Health 2004; 9(4):433-437.

CHAPTER 4

Immune System Polymorphism:
Implications for Genetic Engineering

Tom J. Little*

Summary

As is apparent from the evolution of antibiotic resistance or vaccine escape mutants, parasites and pathogens have the capacity to repeatedly evolve adaptations which enable them to overcome medical interventions. However, evolution may also occur as a natural, ongoing coevolutionary process. Genes involved in the coevolutionary process tend to show high levels of sequence polymorphism because they are the focus of repeated host adaptation and parasite counter adaptation. Knowledge of variation at genes involved in this process, i.e., knowledge of genes that are locked into arms races and thus periodically stimulate pathogen evolution, would seem to be a crucial part of strategies to genetically engineer disease-carrying mosquitoes. Here I summarise what is known about polymorphism in the mosquito immune system and highlight how polymorphism can impact our attempts at intervention. Studies of mosquito immune gene variation are in their infancy, but application of the tools of evolutionary biology holds promise for making the genetic modification of vectors a predictive process.

Introduction

Parasites and pathogens have the capacity to repeatedly evolve adaptations which enable them to overcome medical interventions designed to thwart disease. This is apparent in bacterial antibiotic resistance, vaccine escape mutants, and possibly the evolution of virulence in response to vaccines that immunise relatively weakly.[1] In some cases, it seems that the pathogen's capacity to circumvent medical technology is greater than our capacity (or will) to develop new strategies.[2-4] It is crucial that we learn everything possible from these examples as medical intervention encompasses new technologies such as genetic modification of disease vectors.

However, it is not just medical intervention that drives pathogen evolution. Parasites (by their nature) reduce host survival and reproduction which hosts counter with responses (via their immune system) that in turn reduce parasite fitness. This reciprocal antagonism may lock host and parasite into a process of constant change. Thus, the evolution of pathogens and parasites may also occur as an ongoing, natural coevolutionary process. A significant body of theory supports this: dynamic polymorphisms or arms races are a common outcome of computer simulations of host-parasite coevolution.[5-7] Evolution in response to medical intervention will be influenced by some of the same factors which govern natural coevolutionary dynamics. One prediction from coevolutionary theory is that genes involved in antagonistic interactions will show high levels of DNA sequence polymorphism, and this prediction appears to be met.[8] Framed in reverse, genes which show high levels of adaptive polymorphisms

*Tom J. Little—Institute of Evolutionary Biology, School of Biology, University of Edinburgh, Kings Buildings, West Mains Rd, Edinburgh, EH9 3JT, Scotland, U.K. Email: tom.little@ed.ac.uk

Genetically Modified Mosquitoes for Malaria Control, edited by Christophe Boëte.
©2006 Landes Bioscience.

Table 1. Classification of the genes of the mosquito immune system into four functional classes. N is the number of gene copies present in Anophleles gambiae

Function	Gene	N	Function or Putative Function
Recognition	PGRP	3	Recognises peptidoglycans on pathogen cell surfaces[24]
	TEP	15	Complement-related opsonin[13,57,72]
	GNBP	6	Recognition of gram-negative bacteria, LPS, β-1-3 Glucans[13,73]
	Scavenger Receptor	22	Recognises various ligands, disposes of bacteria.
	C-type Lectins	22	Induced by bacteria and *Plasmodium*, involved in cell-adhesion[74]
Modulation	CLIP-domain serine proteases	41	Associated with TOLL and PO cascade[13,57]
	Serpin	10	Protease inhibitor, upregulated during *Plasmodium* invasion[13,57]
Transduction	Toll/Toll-related	10	Receptor, stimulates cascade for antimicrobials[13,75]
	Relish	2	Transcription factor in Toll cascade[76]
	MyD88	1	Signal transduction in Toll cascade[13]
	Tube	1	Signal transduction in Toll cascade[13]
	Pele	1	Signal transduction in Toll cascade[13]
	Cactus	1	Signal transduction in Toll cascade[13]
	Imd	1	Receptor, stimulates cascade for antimicrobials[13,75]
	STAT	1	Receptor, stimulates cascade for antimicrobials[13,75]
Effectors	Defensin	4	Antimicrobial peptide[13,77]
	Gambicin	1	Antimicrobial peptide[78]
	ICIHT	1	Chitin binding antimicrobial[74]
	Cecropins	4	Antimicrobial peptide[13]
	Prophenol Oxidase	9	Critical for melanin production[13]
	Caspases	12	Implicated in immunity during *Plasmodium* invasion[79]
	Nitric Oxide Synthase	1	NO inhibits parasite development[74]

are likely to be those involved in the coevolutionary process; they are the host defense genes for which parasites have the capacity to adapt against. Such a process brings about the possibility of specific interactions, where host defences are finely tuned to particular pathogen types or strains.[9,10] Similarly, pathogen adaptation to a specific genotypes, is often accompanied by a loss of adaptation to other genotypes.[11]

Thus, understanding of variation at genes involved in resistance would seem to be a crucial part of strategies to genetically engineer disease-carrying mosquitoes. For example, defense genes for which parasites may evolve to overcome are not desirable targets for genetic engineering. In addition, identifying genes involved in resistance polymorphism should enhance our general understanding of parasite strategies to overcome factors (natural or otherwise) which oppose their establishment and development. With a thorough understanding of mosquito immune-gene polymorphisms, it may be possible to make generalisations about the type of host defense gene that malaria is most commonly able to overcome. For example, large amounts of amino acid polymorphism in antimicrobial peptides could indicate that any given variant of these cytosolic peptides has a short evolutionary lifespan of effectiveness, and must constantly evolve to remain effective.

In this chapter I will explore these issues with special reference to the immune system of *Anopheles gambiae*, a species which is the target of genetic engineering to reduce its vectorial capacity with the agent of human malaria, *Plasmodium falciparum*. I need to first summarise

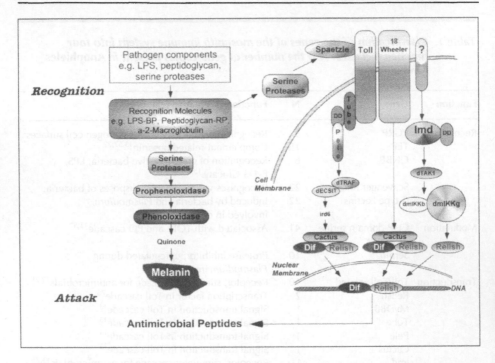

Figure 1. Schematic showing components of the invertebrate innate immune system. Immunity is accomplished through a variety of pattern recognition receptors that detect pathogen molecular signatures and then initiate cascades that ultimately produce products that are harmful to pathogens.

what is know about the mosquito immune system, followed by a description of common analyses of polymorphism aimed at readers unfamiliar with the study of variation. Lastly, I will try to place my ideas in context, and provide examples where evolutionary biology, which is essentially the study of polymorphism, has proven useful in more applied fields and should be incorporated into modification programs.

The Mosquito/Dipteran Immune System

In the past 10 years understanding of the genes controlling the innate immune system of invertebrates has exploded, paralleling a rise in interest in innate immune systems generally.[12] Appreciation of invertebrate immune systems has been advanced largely through the *Drosophila* and *Anopheles* genome projects and associated post-genomic studies.[13,14] However, studies of other organisms have also made notable contributions[15-18] and it is now apparent that virtually all metazoans share certain immune system components.[19,20]

Following Christophides et al (2002), the gene families of the mosquito immune system can be split into four main components: (1) those involved in pathogen recognition, (2) those involved in signal modulation, (3) genes of signal transduction pathways, and (4) effector molecules. A schematic of the humoral immune system is provided (Fig. 1), and a more comprehensive summary of immune-related genes identified in *Anopheles* is given in Table 1. Briefly, recognition molecules (called pattern recognition receptors or PRRs) detect pathogen-associated molecular patterns (PAMPs, typically conserved pathogen cell-surface motifs).[21] Carbohydrates such as lipopolysacharides or peptidoglycans are common PAMPs, though there are a great many other pathogen components known to be stimulants. Through a variety of signal modulation and transduction peptides, PRR's initiate enzyme cascades which ultimately produce the effector molecules that are deadly to pathogens.

```
                    Purifying selection

                    ATG GAA CCA GTG CAT GGT
    Species 1       ATG GAA CCA GTG CAT GGT
                     M   E   P   V   H   G

                    ATG GAG CCA GTC CAT GGG
    Species 2       ATG GAG CCA GTC CAT GGG
                     M   E   P   V   H   G

                    Positive selection

                    ATG GAA CCA GTG CAT GGT
    Species 1       ATG GAA CCA GTG CAT GGT
                     M   E   P   V   H   G

                    ATG AAA CCA ATG CAT CCT
    Species 2       ATG AAA CCA ATG CAT CCT
                     M   K   P   M   H   P

                    Balancing selection

                    ATG AAA CCA GTG CAT CCA
    Species 1       ATG GAA CCA GTT CAT GGT
                     M  E/K  P   V   H  G/P

                    ATG GAT CCA GTC CAT GGT
    Species 2       ATG GAA CAA GTG CAT GAG
                     M  E/D  P   V   H  G/D
```

Figure 2. Illustration of three types of polymorphism that can result from natural selection. In each comparison of two diploid species, there is nucleotide variation in the second, fourth and sixth triplets. For purifying selection, differences among species are largely at silent sites and there is little polymorphism within species. For positive selection (the signature of a molecular arms race) replacement substitutions figure prominently in differences among species, but there still will be little polymorphism within species. The hallmark of balancing selection is polymorphism within populations where alleles may be highly divergent.

Two of these enzyme cascades are particularly well-studied. First, the phenoloxidase cascade produces melanin which is both toxic to pathogens (including Plasmodium[22]) and is used to encapsulate parasitoids.[18] The second notable enzyme cascade is that mediated by membrane-bound TOLL receptors.[23] TOLL or TOLL-like receptors (TLR's) are a conserved component of the innate immune system, present in insects and humans. In insects, it appears that peptidoglycan recognising proteins are one of the important PRR's alerting TOLL's to the presence of invaders. The end products of the intra-cellular cascades originating with TOLL's are cytosolic antimicrobial peptides (effectors). Almost 100 antimicrobial peptides have been identified from insects, and they often occur as multi-gene clusters, function in different ways, and are expressed in different tissues. TOLL's themselves exist as multi-gene families (ten in Anopheles). Moreover, there is diversity in the peptidoglycan-recognising proteins that interact with TOLL,[24,25] thus this cascade and its associated molecules encompass a considerable amount of diversity.

Evolutionary biologists ask a variety of questions about the evolution of this immune-related genome. Because of the conserved nature of innate immunity, analyses of invertebrate systems can tell us something about (1) the origins of the adaptive immune system, even though many of the molecules involved in vertebrate acquired immunity are clearly lacking from invertebrates, or (2) the role of innate immunity in vertebrates. Moreover, because flies have an immune system that is so similar to the vertebrate innate system, this provides the opportunity to study the effects on innate immunity on its own, without the confounding impact of acquired immunity. I will now discuss how and why evolutionary biologists examine polymorphism at particular genes. These analyses are just beginning to be applied to the immune-related genome of invertebrates.

The Study of Sequence Polymorphism

Different forms of pathogen-mediated natural selection leave a distinct stamp on gene sequences, and these differences are discernible when comparing DNA polymorphism patterns among populations and species (Fig. 2).[26-28] For example, with purifying selection, new mutations are less fit and are pruned from the population, thus this process will generate low levels of amino acid polymorphism in genes. Consequently, sequence polymorphism is primarily of the nonsynonymous or silent variety. By contrast, host-parasite coevolution can result in *diversifying* selection which promotes variation for resistance. Diversifying selection, broadly defined, may take two forms. Firstly, host-pathogen interactions can result in arms races, whereby new variants have an advantage and so natural selection proceeds as a series of directional selective sweeps. This is evident as an elevated rate of amino acid replacement among species accompanied by a loss of heterozygosity within species (because mutants tend to go to fixation). Secondly, coevolution may promote diversity by maintaining allelic variants through frequency-dependent or over-dominant selection. The maintenance of polymorphism through these mechanisms is evident as the deep divergence of alleles at single loci, as has occurred for MHC alleles.[29-33] Arms races or balancing selection may act simultaneously.[34]

Variation in infection rates and vectorial capacity may be attributable to genetic variation arising or maintained through these forms of diversifying selection, and it is through molecular population genetic analysis of species and populations that immunity genes associated with coevolutionary diversification are identified. Work of this nature can test the general hypothesis that different genes will be subject to different, predictable forms of selection (e.g., purifying vs directional selection) related to their function. Such work may also illuminate the level of specificity and attenuation typical of interactions. However, the importance of genetic polymorphism in the immune-related genome of mosquitoes, or insects generally, is not yet clear. Compared to other parts of the genome, adaptive polymorphism may be common in insect immune system genes,[35] but much work remains to be done. Indeed, the study of immune-gene polymorphism in arthropods lags behind that of vertebrates or plants.[29-34]

Among the arthropoda, only immune genes from *Drosophila* and the crustacean *Daphnia* have been subject to molecular population genetic analyses of polymorphism. A genome-wide study comparing *D. melanogaster* to *D. simulans* indicated that immune system genes are subject to positive selection to a greater extent than are other parts of the genome.[35] When particular genes have been the target of study in *Drosophila*, results have been mixed: genes for antimicrobial peptides and *Peptidoglycan Recognizing Proteins* largely showed evidence of purifying selection,[36-40] but the transcription factor *Relish* gave evidence of positive selection.[41] Studies of *Daphnia* concerned two genes, one of which, a *Gram Negative Binding Protein* gene showed evidence of purifying selection, while another gene, an *Alpha-2-Macroglobulin*, showed evidence of positive selection.[42] Initial work I have done on *Anopheles* mirrored these results from *Daphnia*. Specifically, an immune related gene in the peptidoglycan recognising group showed evidence of purifying selection, while a Thioester-containing protein (*Alpha-2-Macroglobulin* is within the Thioester-containing family of proteins) showed evidence of positive selection (Little and Cobbe, submitted).

Polymorphism and the Genetic Basis of Resistance in the Mosquito

For many parasitic interactions, it has been possible to identify genes that underlay variation in host susceptibility, and indeed a number of prominent examples come from human interactions with *Plasmodium falciparum*.[31,43,44] However, for most interactions it has not been possible to identify such genes. Based on phenotypic studies, mosquito-pathogen interactions, as with most hosts and pathogens (reviewed in ref. 45), show genetic-based variation for parasite resistance.[46-49] Recently, mosquito genes that crucially mediate *Plasmodium* invasion in mosquitoes have been identified. In particular gene silencing *Anopheles gambiae* C-type lectins greatly enhanced melanisation of *Plasmodium berghei*, while silencing of a leucine rich-repeat protein and a TEP gene greatly inhibited the host immune response.[50,51] Given the key role that these host genes play during *Plasmodium* invasion, one would predict that their action would stimulate an evolutionary response in *Plasmodium* populations, i.e., such genes could be part of an arms race, and this ought to be reflected in patterns of polymorphism.

Overall, however, there is almost no knowledge of polymorphism in mosquito immune systems. At present we can only speculate over which components of the immune system are likely to be the source of variation among mosquitoes By analogy with plant and vertebrate systems, host proteins which recognise pathogens and/or directly interact with pathogens are prime candidates for the detection of adaptive polymorphism.[30,31] However, initial studies on other arthropods do not indicate that this will be the case, as both recognition and attack molecules showed little evidence of adaptive polymorphism (references above).

In regards to recognition molecules, it may be important to consider the various ways in which they work. An important concept in the current understanding of innate immunity is that of the PAMP-PRR interaction.[21,52] Many PAMPs are conserved molecules, often polysaccharides, that are essential for the survival of the pathogen, and as such cannot easily be modified to conceal their recognition by the host. If PAMP escape mutants are unlikely, then PRRs are also unlikely candidates for an arms race. The invertebrate PRR's that recognise conserved polysaccharides probably showed low levels of variation for this reason.[40,42] *Thioester* containing proteins, by contrast, may function as serine protease inhibitors that recognise and bind pathogen proteins. Little (2004) argued that *TEP's* may show evidence of positive selection because they are subject to a host parasite arms race centred on host evolution to produce *TEPs* that inhibit parasite serine proteases and parasite evolution to produce serine proteases that go unrecognised by hosts. TEP genes seem particularly promising targets for the study of polymorphism given their established relevance for vectorial capacity.[51]

In general, gene products involved in protein-protein interactions (as opposed to, for example, the protein-carbohydrate interactions typified by PGRPs or GNBPs) seem more promising candidates for arms races. Serpins, which are common to most immune systems, provide an interesting example of elevated amino acid evolution based on studies of mammals and of parasitic nematodes.[53-55] Host Serpins may function similarly to TEP proteins by binding pathogen serine proteases,[55,56] or they may regulate host serine proteases involved in immune cascades.[18,57] Given this latter function, adaptive evolution of host Serpins might suggest that these proteins are involved in arms races linked to manipulation strategies by pathogens, which will evade the immune response when Serpins are prevented from performing their usual role in the immune response. Another example of an arthropod immune gene showing evidence of elevated amino acid replacement comes from a transcription factor; the NF-κB/IκB protein Relish from *Drosophila*. Indeed, most examples of adaptive polymorphism seem to come from signal modulation or signal transduction genes,[35,38] leading to speculation that most coevolution between insects and their pathogens is centred on immuno-manipulation strategies by pathogens.[41] Concerning malaria/ mosquito interactions it has been suggested in two complementary experiments[58,59] that malaria parasites *P. gallinaceum* are able to suppress *Aedes aegypti* melanisation responses.

Polymorphism and Genetic Engineering

Determining the nature of nucleotide polymorphism in mosquito genes that are relevant for *Plasmodium* invasion would identify genes that are locked into arms races and thus periodically stimulate pathogen evolution. Thus, understanding of polymorphism in *Plasmodium-Anopheles* interactions can aid genetic engineering strategies by helping to determine to types of genes suitable for modification. Unfortunately, the required studies of mosquito immune gene variation are in their infancy. There are some analogies here with the challenges faced when choosing antigens for vaccine development in vertebrates. On one hand, it is sensible to choose antigens that that the immune system has the capacity to detect, but these tend to show large amounts of polymorphsim, which is an adaptation against immune defense. Antigens based on polymorphic proteins may not give widespread protection if strain variation in the wild is beyond the degeneracy of the immune response to the vaccinating antigen. The issue is with the degree of specificity. Highly invariant proteins seem better targets for vaccine development, but there conservation often indicates that host immune system is not sensitive to them.

To further facilitate predictions about the likely outcome of parasite evolution on genetically modified mosquitoes, it should be possible to study the experimental evolution of *Plasmodium* through natural and modified mosquitoes, and has been accomplished with parasite passages through vertebrate hosts.[60] As part of experimental studies of polymorphism, it will be necessary to establish the level of specificity and attenuation. Given the possibility of highly specific interactions,[46,61] it may be necessary to avoid model systems and perform experiments on natural combinations of the species of greatest medical relevance.[62] With deep knowledge of both natural levels of DNA polymorphism and likely adaptive outcomes established through experimental evolution, it may be possible to make robust predictions regarding the spread of strains resistant to particular modifications of mosquito defenses. Naturally, a number of outcomes will be difficult to predict. For example, if any introduced gene is foreign, i.e., has never naturally been part of a mosquito genome, it would be very hard to predict its performance within its new host genome.[63] Or, if the introduction of a genetically modified mosquito stimulates pathogen adaptation against the newly introduced allele, this could relax selection on other parts of the immune-related genome that would otherwise be under parasite-mediated selection.

An appropriate comparison is with the evolution of antibiotic resistance, where there is now thorough knowledge of both the mechanisms that confer resistance and their energetic costs.[64] The study of antibiotic resistance has advanced to the stage where predictions about the speed of evolution are feasible, including the number of amino acid substitutions required to confer resistance.[65] Of course, work on antibiotic resistance was greatly accelerated by its rise to prominence as a serious medical health issue. We should attempt to avoid a similar progression of events and *post hoc* problem solving regarding *Plasmodium* evolution in response to genetically engineered mosquitoes.

In general, understanding the capacity of parasite or pathogens to overcome host defenses or medical interventions ultimately requires the tools of evolutionary biology. In addition to the arguments made above, the tools of evolutionary biology and ecology may also contribute to the effective use of genetically engineered mosquitoes through the study of trade-offs,[66,67] genotype x environment interactions,[68-70] or maternal effects,[71] all of which greatly influence host-parasite interactions. Given such a range of factors that influence the evolutionary success of a genome, my suspicion is that the successful introduction of a malaria combating mosquito stands as a formidable challenge. Even if this laudable goal is not achieved, the use of the tools of evolutionary biology in pursuit of this goal still offers rewards because of the gain in fundamental understanding of invertebrate immune systems.

References

1. Read AF, Gandon S, Nee S et al. Infectious Disease and Host-pathogen Evolution. In: Dronamraju KR, ed. Cambridge: Cambridge University Press, 2004:265-292.
2. Cohen ML. Changing patterns of infectious disease. Nature 2000; 406:762-767.
3. Gilmore MS, Hoch JA. Antibiotic resistance - A vancomycin surprise. Nature 1999; 10;399(6736):524-5, 527.
4. Novak R, Henriques B, Charpentier E et al. Emergence of vancomycin tolerance in streptococcus pneumoniae. Nature 1999; 399:590-593.
5. Hamilton WD. Sex versus nonsex versus parasite. Oikos 1980; 35:282-290.
6. Anderson RM, May RM. Coevolution of hosts and parasites. Parasitology 1982; 85:411-426.

7. Seger J. Dynamics of some simple host-parasite models with more than two genotypes in each species. Phil Trans Roy Soc B319 1988:541-555.

8. Hurst LD, Smith NGC. Do essential genes evolve slowly? Curr Biol 1999; 9:747-750.

9. Schmid-Hempel P, Ebert D. On the evolutionary ecology of specific immune defence. Trends Ecol Evol 2003; 18:27-32.

10. Carius H-J, Little TJ, Ebert D. Genetic variation in a host–parasite association: Potential for co-evolution and frequency dependent selection. Evolution 2001; 55:1136-1145.

11. Ebert D. Experimental evolution of parasites. Science 1998; 282:1432-1435.

12. Kurtz J. Memory in the innate and adaptive immune systems. Microbes Infect 2004; 6:1410-1417.

13. Christophides GK, Zdobnov E, Barillas-Mury C et al. Immunity-related genes and gene families in Anopheles gambiae. Science 2002; 298:159-165.

14. Christophides GK, Vlachou D, Kafatos FC. Comparative and functional genomics of the innate immune system in the malaria vector Anopheles gambiae. Immunoll Rev 2004; 198:127-148.

15. Bachere E, Destoumieux D, Bulet P. Penaeidins, antimicrobial peptides of shrimp: A comparison with other effectors of innate immunity. Aquaculture 2000; 191:71-88.

16. Iwanaga S. The molecular basis of innate immunity in the horseshoe crab. Curr Opin Immunol 2002; 14:87-95.

17. Ochiai M, Ashida M. A pattern recognition protein for peptidoglycan- cloning the cDNA and the gene of the silkworm, Bombyx mori. J Biol Chem 1999; 274:11854-11858.

18. Soderhall K, Cerenius L. Role of the prophenoloxidase-activating system in invertebrate immunity. Curr Opin Immunol 1998; 10:23-28.

19. Hoffmann JA, Kafatos FC, Janeway CA et al. Phylogenetic perspectives in innate immunity. Science 1999:1313-1318.

20. Dangle JL, Jones JDG. Plant pathogens and integrated defence in plants. Nature 2001; 411:826-833.

21. Medzhitov R, Janeway CA. Innate immunity: The virtues of a nonclonal system of recognition. Cell 1997; 91:295-298.

22. Collins FH, Sakai RK, Vernick KD et al. Genetic selection of a Plasmodium-refractory strain of the malaria vector Anopheles-gambiae. Science 1986; 234:607-610.

23. Aderem A, Ulevitch RJ. Toll-like receptors in the induction of the innate immune response. Nature 2000; 406:782-787.

24. Michel T, Reichhart J-M, Hoffmann JA et al. Drosophila toll is activated by Gram-positive bacteria through a circulating peptidoglycan recognition protein. Nature 2001; 414:756-759.

25. Choe KM, Werner T, Stoven S et al. Requirement for a peptidoglycan recognition protein (PGRP) in relish activation and antibacterial immune responses in Drosophila. Science 2002; 296:359-362.

26. Ford MJ. Applications of selective neutrality tests to molecular ecology. Mol Ecol 2002; 11:1245-1262.

27. Olson S. Seeking the signs of selection. Science 2002; 298:1324-1325.

28. Yang Z, Bielawski JP. Statistical methods for detecting molecular adaptation. Trends Ecol Evol 2000; 15:496-502.

29. Stahl EA, Dwyer G, Mauricio R et al. Dynamics of disease resistance polymorphism at the Rpm1 locus of Arabidopsis. Nature 1999; 400:667-671.

30. Stahl EA, Bishop JG. Plant-pathogen arms races at the molecular level. Curr Opin Plant Biol 2000; 3:299-304.

31. Hill AVS, Allsopp CEM, Kwiatkowski D et al. Common West African HLA antigens are associated with protection from severe malaria. Nature 1991; 352:595-600.

32. Hill AVS, Kwiatkowski D, McMichael AJ et al. Maintenance of Mhc Polymorphism - Reply. Nature 1992; 355:403-403.

33. Hughes AL, Nei M. Maintenance of Mhc Polymorphism. Nature 1992; 355:402-403.

34. Bergelson J, Kreitman M, Stahl EA et al. Evolutionary dynamics of plant R-genes. Science 2001; 292:2281-2285.

35. Schlenke TA, Begun DJ. Natural selection drives Drosophila immune system evolution. Genetics 2003; 164:1471-1480.

36. Ramos-Onsins S, Aguade M. Molecular evolution of the Cecropin multigene family in Drosophila: Functional genes vs. pseudogenes. Genetics 1998; 150:157-171.

37. Clark AG, Wang L. Molecular population genetics of Drosophila immune system genes. Genetics 1997; 147:713-724.

38. Lazzaro BP, Clark AG. Molecular population genetics of inducible antibacterial peptide genes in Drosophila melanogaster. Mol Biol Evol 2003; 20:914-923.

39. Date A, Satta Y, Takahata N et al. Evolutionary history and mechanism of the Drosophila cecropin gene family. Immunogenetics 1998; 47:417-429.

40. Jiggins FM, Hurst GDD. The evolution of parasite recognition genes in the innate immune system: Purifying selection on Drosophila melanogaster peptidoglycan recognition proteins. J Mol Evol 2003; 57:598-605.
41. Begun DJ, Whitley P. Adaptive evolution of relish, a Drosophila NF-kappa B/I kappa B protein. Genetics 2000; 154:1231-1238.
42. Little TJ, Colbourne JK, Crease TJ. Molecular evolution of Daphnia immunity genes: Polymorphism in a gram negative binding protein and an Alpha-2-macroglobulin. J Mol Evol 2004; 59:498-506.
43. Ntoumi F, Rogier C, Dieye A et al. Imbalanced distribution of Plasmodium falciparum MSP-1 genotypes related to sickle-cell trait. Mol Med 1997; 3:581-592.
44. Ruwende C, Khoo SC, Snow RW et al. Natural selection of hemi- and heterozygotes for G6PD deficiency in Africa by resistance to severe malaria. Nature (London) 1995; 376:246-249.
45. Little TJ. The evolutionary significance of parasitism: Do parasite-driven genetic dynamics occur ex silico? J Evol Biol 2002; 15:1-9.
46. Ferguson HM, Read AF. Genetic and environmental determinants of malaria parasite virulence in mosquitoes. Proceedings of the Royal Society of London Series B-Biological Sciences 2002; 269:1217-1224.
47. Hogg JC, Hurd H. Malaria-induced reduction of fecundity during the first gonotrophic cycle of Anopheles-stephensi mosquitos. Med Vet Entomol 1995; 9:176-180.
48. Yan G, Christensen BM, Severson DW. Comparisons of genetic variability and genome structure among mosquito strains selected for refractoriness to a malaria parasite. J Hered 1997; 88:187-194.
49. Bosio CF, Fulton RE, Salasek ML et al. Quantitative trait loci that control vector competence for dengue-2 virus in the mosquito Aedes aegypti. Genetics 2000; 156:687-698.
50. Osta MA, Christophides GK, Kafatos FC. Effects of mosquito genes on Plasmosium development. Science 2004; 303:2030-2032.
51. Blandin S, Shiao SH, Moita LF et al. Complement-like protein TEP1 is a determinant of vectorial capacity in the malaria vector Anopheles gambiae. Cell 2004; 116:661-670.
52. Janeway CA, Medzhitov R. Innate immune recognition. Ann Rev Immunol 2002; 20:197-216.
53. Gill DE, Mock BA. Ecology and genetics of Host-parasite interactions. In: Rollison D, Anderson RM, eds. The Linnean Society of London 1985:157-183.
54. Zang XX, Maizels RM. Serine proteinase inhibitors from nematodes and the arms race between host and pathogen. Trends Biochem Sci 2001; 26:191-197.
55. Barbour KW, Goodwin RL, Guillonneau F et al. Functional diversification during evolution of the murine alpha(1)-proteinase inhibitor family: Role of the hypervariable reactive center loop. Mol Biol Evol 2002; 19:718-727.
56. Kanost MR. Serine proteinase inhibitors in arthropod immunity. Dev Comp Immunol 1999; 23:291-301.
57. Oduol F, Xu JN, Niare O et al. Genes identified by an expression screen of the vector mosquito Anopheles gambiae display differential molecular immune response to malaria parasites and bacteria. Proc Natl Acad Sci USA 2000; 97:11397-11402.
58. Boëte C, Paul REL, Koella JC. Reduced efficacy of the immune melanization response in mosquitoes infected by malaria parasites. Parasitology 2002; 125:93-98.
59. Boëte C, Paul REL, Koella JC. Direct and indirect immunosuppression by a malaria parasite in its mosquito vector. Proceedings of the Royal Society of London Series B-Biological Sciences 2004; 271:1611-1615.
60. Mackinnon MJ, Read AF. Immunity promotes virulence evolution in a malaria model. Plos Biology 2004; 2:1286-1292.
61. Lambrechts L HJ, Durand P, Gouagna LC et al. Host genotype by parasite genotype interactions underlying the resistance of anopheline mosquitoes to Plasmodium falciparum. Malari J 2005; 4(1):3.
62. Tahar R, Boudin C, Thiery I et al. Immune response of Anopheles gambiae to the early sporogonic stages of the human malaria parasite Plasmodium falciparum. EMBO J 2002; 21:6673-6680.
63. Moreira LA, Wang J, Collins FH et al. Fitness of anopheline mosquitoes expressing transgenes that inhibit plasmodium development. Genetics 2004; 166:1337-1341.
64. Baquero F, Blazquez J. Evolution of antibiotic resistance. Trends Ecol Evol 1997; 12:482-487.
65. Barlow M, Hall BG. Experimental prediction of the natural evolution of antibiotic resistance. Genetics 2003; 163:1237-1241.
66. Ebert D, Carius H-J, Little TJ et al. The evolution of virulence when parasites cause host castration and gigantism. American Naturalist 2004; 164 Suppl 5:S19-32.
67. Lenski RE, Souza V, Duong LP et al. Epistatic effects of promoter and repressor functions of the Tn10 tetracycline-resistance operon on the fitness of Escherichia coli. Mol Ecol 1994; 3:127-135.

68. Mitchell SE, Rogers ES, Little TJ et al. Host-parasite and genotype-by-environment interactions: Temperature modifies potential for selection by a sterilizing pathogen. Evolution 2005; 59:70-80.

69. Stacey DA, Thomas MB, Blanford S et al. Genotype and temperature influence pea aphid resistance to a fungal entomopathogen. Physiol Entomol 2003; 28:75-81.

70. Blanford S, Thomas MB, Pugh C et al. Temperature checks the Red Queen? Resistance and virulence in a fluctuating environment. Ecol Lett 2003; 6:2-5.

71. Little TJ, O'Connor B, Colegrave N et al. Maternal transfer of strain-specific immunity in an invertebrate. Curr Biol 2003; 13:489-492.

72. Levashina EA, Moita LF, Blandin S et al. Conserved role of a complement-like protein in phagocytosis revealed by dsRNA knockout in cultured cells of the mosquito, Anopheles gambiae. Cell 2001; 104:709-718.

73. Dimopoulos G, Richman A, Muller HM et al. Molecular immune responses of the mosquito Anopheles gambiae to bacteria and malaria parasites. Proc Natl Acad Sci USA 1997; 94:11508-11513.

74. Dimopoulos G, Seeley D, Wolf A et al. Malaria infection of the mosquito Anopheles gambiae activates immune-responsive genes during critical transition stages of the parasite life cycle. EMBO J 1998; 17:6115-6123.

75. Luo CH, Zheng LB. Independent evolution of Toll and related genes in insects and mammals. Immunogenetics 2000; 51:92-98.

76. Shin SW, Kokoza V, Ahmed A et al. Characterization of three alternatively spliced isoforms of the Rel/NF-kappa B transcription factor Relish from the mosquito Aedes aegypti. Proc Natl Acad Sci USA 2002; 99:9978-9983.

77. Chalk R AC, Ham PJ, Townson H. Full sequence and characterization of two insect defensins: Immune peptides from the mosquito Aedes aegypti. Proc Royal Soc London - Series B: Biological Sciences 1995; 261:217-221.

78. Vizioli J, Bulet P, Hoffmann JA et al. Gambicin: A novel immune responsive antimicrobial peptide from the malaria vector Anopheles gambiae. Proc Natl Acad Sci USA 2001; 98:12630-12635.

79. Han YS, Thompson J, Kafatos FC et al. Molecular interactions between Anopheles stephensi midgut cells and Plasmodium berghei: The time bomb theory of ookinete invasion of mosquitoes. EMBO J 2000; 19:6030-6040.

CHAPTER 5

Predicting the Spread of a Transgene in African Malaria Vector Populations:
Current Knowledge and Limitations

Frédéric Simard* and Tovi Lehmann

Abstract

One strategy for the control of malaria and other vector-borne diseases relies on the ambitious goal of depleting natural vector populations' ability to transmit the pathogen through the introduction and spread of an engineered genetic construct. In this chapter, we assess whether the data accumulated so far on the population genetic structure of *Anopheles gambiae*, the major human malaria vector in Africa and the one studied most extensively, can be used to predict the spread of such genetic construct within and between wild populations. We conclude that available data offer good qualitative description of *An. gambiae* population structure, but do not provide the necessary information on the processes shaping this structure. We explore biological and methodological issues that prevented derivation of quantitative descriptions of these processes, focusing on the estimation of the effective population size and gene flow between populations. We discuss plans for bridging the gap between our present knowledge and where we should be, and outline a protocol for the direct estimation of relevant population genetics parameters and quantitative assessment of their interaction through a field population perturbation study. Finally, the epidemiological importance of other vector species in sustaining malaria transmission is highlighted as an additional roadblock that needs to be considered as part of any comprehensive vector control strategy designed to substantially lower the burden of malaria that overwhelms Africa.

Introduction

Novel approaches for the control of malaria transmission through genetic alteration of their mosquito vectors have received considerable attention in the past decade.[1,2] They rely on the effective spread of transgene(s), i.e., gene(s) engineered to reduce vector competence such as by conferring refractoriness against the parasite,[3-5] within natural vector mosquito populations. This suggests that the basis of the control, e.g., the transgene(s), will first be introduced (artificially) into the natural vector population(s) and that it will subsequently be transmitted to the offspring, to the extent that, within several generations, practically all

---

*Corresponding Author: Frédéric Simard—Institut de Recherche pour le Développement (IRD), Unité de Recherche 016 "Caractérisation et contrôle des populations de vecteurs", Laboratoire de Recherche sur le Paludisme, Organisation de Coordination pour la lutte Contre les Endémies en Afrique Centrale (OCEAC), BP 288, Yaoundé, Cameroun. Email: simard@ird.fr. Website: www.mpl.ird.fr/ur016/

Genetically Modified Mosquitoes for Malaria Control, edited by Christophe Boëte.
©2006 Landes Bioscience.

individuals of the target species will express the refractory phenotype. Genetic drive mechanisms that should speed-up the process or improve efficient heritability are being developed and have received at least proof of principle.[6-8] However, one fundamental assumption of this strategy is that mating occurs between individuals that carry the transgene(s) and individuals that do not.[9] In other words, the target vector population is assumed to be a single, randomly mating unit, whereas assortative mating in wild mosquito populations has been demonstrated and could affect the spread of a transgene.[10-14] Successful spread of such transgenes therefore depends on our ability to describe the basic reproductive units (demes, see Box 1) that compose the vector system responsible for malaria transmission in Africa, to understand their genetic and population dynamics, and determine the forces that shape it. Lessons from past genetic control programs demonstrated that the population structure and population dynamics of the target population(s) determine which, if any, genetic control approaches would be appropriate for addressing a specific problem.[15] A critical part of this is obtaining a quantitative understanding of the spatial and temporal population structure of the mosquito vector. Such data are needed as input parameters for constructing predictive models for the prospects of different strategies to introduce genes into these populations. This constitutes the rationale for most population genetics studies aiming at unravelling the genetic structure of African malaria vectors.

In the following, we assess whether the available population genetics knowledge provides a solid basis for predicting the spread of a gene within and among natural malaria vector populations, with an emphasis on *Anopheles gambiae*, the most important vector throughout Africa and the most likely target for genetic control. As such, members of the *An. gambiae* complex have been extensively studied providing the most detailed information on their population biology and genetic structure. A number of reviews have been published recently on the knowledge gathered to date on this species complex (see for example refs. 9,16-22). We do not attempt to duplicate this work. Rather, we assess whether these data can be used to predict the spread of an introduced gene within and between natural *An. gambiae* populations. Finally, we discuss the expected impact of the transgenic approach over malaria transmission in Africa.

Box 1. Glossary of terms

Allele: the state of a gene at a locus that differs from other such alleles by one or more mutations (e.g., DNA sequence differences).

Deme: the local breeding unit of a species within which individuals mate at random and genotype frequencies of neutral alleles are at Hardy-Weinberg equilibrium.

(Random genetic) drift: random change in allele (gene) frequencies that occurs over generations as a result of the finite number of gametes from the parent generation that form the subsequent generation.

Effective population size (*Ne*): a measure of genetic drift that can be approximated as the number of parents that contribute gametes to the next generation within a deme, assuming equal sex ratio and identical reproductive potential.

Gene flow (*Nm*): the spread of a gene or allele as a result of mating between individuals from different populations.

Introgression: gene flow between species by hybridization and backcrossing.

Norm of reaction: the array of phenotypes that a single gene or allele can provide in a range of genetic backgrounds and external environments.

Reciprocal monophyly: an outcome of the stochastic loss of ancestral polymorphism over time in two populations or two species derived from a common ancestral source corresponding to the presence of only unique alleles in each group (species).

Predicting the Spread of a Gene within and among Natural Vector Populations

Implementation of a novel public health control operation on a magnitude of a continent demands the highest and most rigorous preparation.[2,23] The introduction and spread of genes into natural vector populations to interrupt disease transmission cannot be imagined without the capacity to predict, with sufficient accuracy, the outcome of a release effort. Prediction of changes in allele (gene) frequencies over time and space depends upon reasonable estimates of key parameters of the processes that determine such changes. The relevant outcomes are (i) the time until establishment of the introduced gene locally, within a single deme and (ii) the time for the gene establishment in other demes via natural spread. Establishment is defined as fixation (frequency = 1) or the frequency of stable equilibrium for the introduced gene. Such predictions require knowledge of contemporary migration between demes, selection, and drift as well as estimating the key parameters of these processes. Box 2 lists a minimal set of parameters, estimates of which are required to predict future changes in allele frequencies. Although not an exhaustive list, predictions based on fewer parameters may provide questionable results. Estimates of some of these parameters are found in the literature however, they suffer from serious flaws.

The Selective Value of the Transgene

In the absence of a genetic drive mechanism, establishment and further spread of a gene conferring refractoriness to malaria infection in wild mosquito population(s) will essentially rely on its net selective value, i.e., the balance between the fitness cost of phenotypic expression of the (introduced) gene and the overall benefit for the mosquito by escaping the detrimental effect of parasite infection and possibly, protection from other pathogens as well.[24-26] Depending on the gene(s) involved, and their underlying expression dynamics, maintenance costs might be fixed (if the gene is to be expressed constitutively) or conditional (if the gene is expressed in certain conditions, e.g., in response to parasite infection in which case, the evolutionary cost of refractoriness is obviously sex specific, because only female anopheles are exposed to malaria parasites, and a function of the probability that the mosquito becomes infected). Furthermore, it is likely that environmental factors and the genetic background of differentially adapted vector populations will modulate the balance of evolutionary cost and benefit of refractoriness. The norm of reaction of any gene to be introduced in the genome of a vector species therefore needs to be assessed across the range of genetic variability the target species possesses and the diverse environments it experiences to assure that the phenotype (i.e., refractoriness) is predictable.[23,27,28] This is a formidable challenge because relevant parameters of neither the natural environment (temperature, humidity, diet, crowding...) nor the relevant genetic variability (nucleotide polymorphism, genome structure, chromosomal inversions, cytological position...) are clearly defined. Although insights can be gained from cage experiments, whether these are conducted in a laboratory or in semi-field conditions will reflect at best only a parcel of the outcomes expected in a species like *An. gambiae* in nature. In this respect, the analysis of the spread of the *Kdr* mutation conferring insecticide resistance is very appealing because it represents the spread of a new gene under selection in natural settings. This single-nucleotide mutation was originally described from West African *An. gambiae* populations[29] that are known to be genetically and ecologically differentiated subpopulations.[9,16,19] Despite an apparently obvious fitness benefit in areas of intensive insecticide use, the *Kdr* allele was found only in populations of the S molecular form of *An. gambiae* and not in sympatric populations of the M form.[30,31] It was subsequently found in the M form after an apparent introgression event from the S form,[32] and is now spreading in this form as well (Etang J, Fondjo E, Simard F, unpublished).[33,34] In the sibling species, *An. arabiensis*, the *Kdr* mutation apparently emerged as an independent mutation.[35] The actual geographic distribution

---

Box 2. Predicting the spread of a gene in natural malaria vector populations: what do we need to know?

Changes in allele frequency over time and space depend on properties of the allele, the subpopulation, the rates of gene exchange between subpopulations, and the interactions between these properties. Within a breeding unit (deme), the future change in allele frequency depends on its selective value (i.e., its fitness), its initial frequency at introduction, and the deme's effective population size (Ne). Furthermore, planning effective introduction of a gene into the local breeding unit requires having reasonable estimates of the geographical area it encompasses, and of the adult population size. Since the seasonal dynamics of these vector populations generally involves dramatic changes, it will be needed to know the seasonal and spatial dynamics in these parameters. The large differences in the population structure of *An. gambiae* in West and East Africa, and the remarkable environmental heterogeneity across the species range, requires consideration of the difference in these parameters between regions and environments.

Gene and drive system parameters to be known include:
  i. The net selective value of the introduced gene (including its genetic drive system) for uninfected and infected mosquitoes is the main predictor of the systematic change of its frequency over generations,
 ii. The norm of reaction of any candidate gene conferring refractoriness needs to be assessed,
iii. The stability of the transgene construct with respect to recombination and mutations rendering it ineffective needs to be addressed.

Vector populations' parameters:
  i. The effective population size (Ne) of the basic reproductive units (demes) is required to calculate the lowest net selective value the gene should have to overcome probability of loss due to the stochastic variance over generations in allele frequencies (i.e., random genetic drift),
 ii. The corresponding size of the adult population (estimated count) is needed to calculate the allele frequency at introduction,
iii. The geographical area occupied by a deme is required to calculate the number of such units per region.

Gene flow between populations:
  i. Contemporary rates of gene flow between demes separated by distance or other barriers to gene flow will be required to calculate the rate of spread of the gene over space,
 ii. Knowledge of the geographic and biological (pre and post-mating) barriers that prevent or hinder populations' admixture is needed to assess their strength and stability in time,
iii. If gene flow involves "rare events" such as extinction-colonization or accidental migration, the frequency of these events needs to be assessed and their underlying (ecological) causes need to be identified.

---

of the *Kdr* mutation in the *An. gambiae* complex suggests fluctuating balance between evolutionary costs and benefits that might favor its spread under certain ecological conditions only.[32-34] It is likely that similar limitations applies to any mutation or gene with a strong phenotypic effect.

However, in the case of genetically engineered mosquitoes, an efficient drive system should promote the spread of refractoriness allele(s), even in the face of unfavorable balance of evolutionary costs of refractoriness.[24,25] Concerns about the stability of the genetic construct will need to be addressed separately,[23,36,37] and the efficacy of the drive mechanism in promoting the spread of the transgene will need to be demonstrated under a variety of natural conditions.[38] Indeed, although robust inferences were generated from theoretical work,[6,24,25] experimental evidence for efficient drive mechanisms in mosquitoes has yet to be provided.[37-40] Furthermore, as outlined above, the genetic structure of natural vector populations will mediate the spread of genes in space and time.

Limitations in the Knowledge of the Population Structure of African Malaria Vectors

Estimating Effective Population Size

The effective population size (*Ne*) reflects the degree to which a population is affected by random genetic drift.[41] Genetic drift affects the stability of allele frequencies in populations over generations, such that large fluctuations in allele frequencies are expected in small populations, whereas small changes would occur in large populations.[42,43] Hence, genetic drift influences the magnitude of genetic diversity within a population and the rate of differentiation between populations. *Ne* depends on demographic factors such as population density, dispersal, and the mating system. When population size varies among generations, *Ne* approximates the harmonic mean of the effective population sizes in each single generation, and hence is dominated by the smallest value.[44,45] Episodes of small *Ne* (i.e., demographic and genetic bottlenecks) can be of great evolutionary significance because increased genetic drift during these periods can dramatically change allele frequencies and the distribution of the overall genetic variability within and between populations. In particular, a transient drop in *Ne* can favour the rise in frequency of alleles that otherwise would have been selected against due to fitness cost.

Several methods are available to estimate *Ne* from demographic or genetic data. They vary in the types of information they use, their sensitivity to various assumptions, and most importantly, they refer to somewhat different definitions of *Ne*.[46-50] The most widely used genetic estimator derives *Ne* from the variance in allele frequencies between generations.[42,44,51] The method has been used to estimate the effective population size of *An. gambiae* and *An. arabiensis* in a number of settings and using various genetic markers.[43,48,51-54] Reported estimates of *Ne* were in the thousands for both species (but see ref. 53 for a geographically isolated *An. arabiensis* population) and significant differences in *Ne* were demonstrated between populations of *An. gambiae*.[51] Overlooking such differences in *Ne* between populations leads to erroneous estimates of genetic differentiation, gene flow and divergence time.[55,56] However, further interpretation of the results in a quantitative way had to be tentative, because the method relies on assumptions that do not hold true in populations of *An. gambiae*. These assumptions include random mating and equal reproductive potential across individuals, nonoverlapping generations, equal sex-ratio, and negligible selection, migration and mutation. Evidences showing that several of these assumptions are violated in natural vector populations have accumulated. Such violations can lead to severely biased *Ne* estimates. Further, *Ne* estimates derived through such moment-based estimators are biased upward.[49] Often, confidence intervals around the estimated values of *Ne* were so wide that the estimate's biological significance was lost. Hence, although valuable to compare populations to one another, the available estimates of *Ne* are not suitable for use in predictive models of the spread of alleles within and between populations.

New methods are being developed to improve estimation of *Ne*, which appear robust over a wide range of realistic conditions due to relaxed assumptions.[49] However, predicting the spread of an introduced gene in natural *An. gambiae* populations will require a detailed picture of the fluctuations, both in time and space, of the effective population size of target populations. Precise assessment of the number of reproductively active adults in a population is needed to plan the release effort, as well as the identification of the time and place where the natural population is most amenable to the genetic introgression of a new gene or allele. The geographical area associated with a deme and how this area varies in different environments may also be valuable for optimizing the release effort.

The above discussion assumes the existence of discrete demes as the building blocks of *An. gambiae* gene pool. However, some evidence suggests that the gene pool of *An. gambiae* in Africa is divided into few large subdivisions, within which isolation by distance applies.[57-59] Under this model instead of discrete breeding units, there is a continuum where geographically closer populations are genetically more similar and reproductive adults disperse to all

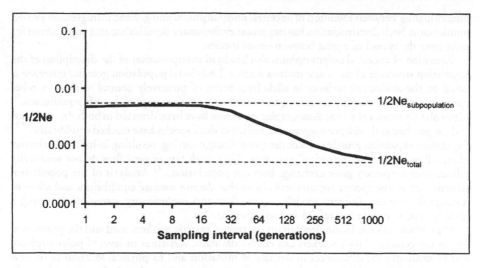

Figure 1. The effect of migration on the estimate of Ne obtained from temporal changes in allelic frequencies using the moment estimator.[42] The total population (Ne_{total} = 1100; $1/2Ne_{total}$ = 0.0045) is fragmented in 11 subpopulations of equal effective population size ($Ne_{subpopulation}$ = 100; $1/2Ne_{subpopulation}$ = 0.05), exchanging migrants in an island model at a rate m = 0.2. As can be seen, when sampling interval is small (i.e., <16 generations in this example), Ne estimates are close to the true value. However, as time between samples taken for the temporal analysis increases, estimates of Ne approach the true value for the whole species. Note that both axes are in log scale. Adapted from Wang J, Whitlock MC; Estimating population size and migration rates from genetic samples over space and time; Genetics 2003; 163:429-446; with permission from the Genetics Society of America.[64]

directions randomly with no barriers, except for their dispersal capacity and the limits of the subdivision or the species range.[60-63] If the isolation by distance model accurately reflects the genetic structure of *An. gambiae*, then the values of Ne obtained so far do not refer to actual demes and are not useful. High rate of migration between populations acts as a buffer against genetic drift and results in estimate of Ne that increases as the period between the samples taken to estimate the variance in allele frequencies is longer, in sharp contrast to expectations if the estimate applies to a single deme (Fig. 1).[64] Such results were obtained for an *An. gambiae* population from western Kenya (Lehmann et al unpublished) and for *An. arabiensis* in Cameroon.[54] More studies are needed to determine if isolation by distance better describes the organization of the gene pool of *An. gambiae* and other malaria vectors in Africa before interpretation of the Ne estimates can be made.

Estimating the Level of Gene Flow between Populations

The principal malaria vectors in Africa (and typically elsewhere) are members of sibling (or cryptic) species complexes.[65] Morphologically, the members of a complex are indistinguishable, reflecting that these species have diverged very recently. Some authors argued that speciation within the *An. gambiae* species complex, and most importantly lineage splitting between *An. arabiensis* and *An. gambiae*, occurred less than 4,000 years ago, as a byproduct of the development of agriculture in formerly unfavorable central African rainforest areas.[16,17,66] As such, these species may retain substantial amounts of shared ancestral polymorphism because insufficient time has elapsed for reciprocal monophyly to establish.[67] Post-mating reproductive barriers between members of the *An. gambiae* complex are incomplete, because only male hybrids are sterile but females are fertile, allowing some genetic exchange. Compelling evidence that such process occurred between *An. arabiensis* and *An. gambiae* in, at least parts of, their genomes, has been provided in experimental as well as natural settings.[68,69] As a result, discriminating between retention of ancestral polymorphism and genetic introgression proved problematic. Such discrimination has important evolutionary significance and implications for estimating the spread of a gene between vector species.

Retention of ancestral polymorphism also hindered interpretation of the description of the population structure of the major malaria vectors. Traditional population genetics inference is based on the analysis of variance in allele frequencies of putatively neutral markers. It relies upon a number of simplifying assumptions such as mutation-migration-drift equilibrium.[70] Molecular signatures of recent demographic expansion have been detected in both *An. arabiensis* and *An. gambiae* and evidence suggests that neither of these species have reached equilibrium.[57,71] Population expansion greatly reduces the rate of lineage sorting, resulting in inflated estimates of gene flow (Nm)[72] by "historical" gene flow. Thus, high rate of gene flow do not necessarily reflects contemporary gene exchange between populations.[73] Analysis of the population structure of vector species requires techniques that do not assume equilibrium and allow to distinguish between different models of gene flow and evolutionary scenarios explaining a given genetic structure (see ref. 74 and refs. therein).[74]

Population differentiation depends on the type of genetic markers used and the position of loci in the genome. DNA markers can exhibit dramatic variations in level of polymorphism due to locus-specific differences in the rate of mutation and to physical location in or near chromosomal inversions or loci under selection (a process that is known as 'genetic hitchhiking').[75] Hence, results obtained from the same species using different types of markers or different sets of loci will not necessarily agree[21,76,77] and summary statistics representing genome-wide trends must exclude outlier/deviant loci.[59] Distinguishing locus-specific from genome-wide effects is a prerequisite for a correct description of population structure. Furthermore, estimates of genetic differentiation between populations depend on the analytic method used and the (evolutionary and demographic) model assumed.[77] As no consensus has yet been reached, comparison across studies remains problematic.

With these limitations, it is not surprising that the low level of genetic divergence typically observed between natural populations of *An. gambiae* led to largely inconclusive results as far as contemporary gene flow is concerned.[19,59,78] Similar finding seems to emerge from recent analyses conducted in the other major human malaria vector, *An. funestus*.[79-81] However, few consistent trends have emerged providing a good qualitative description of the patterns of gene flow between *An. gambiae* populations. In the face of shallow geographical population structure between neighboring populations, recent studies revealed strong, if incomplete, barriers to gene flow between the molecular forms M and S of *An. gambiae*.[18-20,59,82] Because both forms have extensively overlapping geographical and temporal distributions and are widespread throughout the continent,[20] such findings suggest that genes might spread over large geographical areas, within one molecular form, before potentially invading the other form.[14] This is reminiscent of the *Kdr* gene situation described above. However, the degree of differentiation between molecular forms appears very low over most of the genome, but is remarkably high in few small genomic regions not only because of paracentric inversions.[59,77,83,84] Such semipermeable barriers to gene flow in a mosaic genome prompts further studies to identify regions of the genomes with different abilities to introgress between molecular forms and species within the *An. gambiae* complex.

Proposed Plan for Bridging the Gap

Population genetics studies produced robust description of the population structure, but they failed to quantify the processes that have shaped this structure. As stated by Gould and Schliekelman[15] "Researchers working with classical genetic manipulations learned over and over again that there is no substitute for examining behavior of a genetically manipulated strain under local field conditions. This will not change in the future". We echo their view and advocate that, the ultimate approach to estimate contemporary gene flow and derive robust estimates of all key parameters is by tracking new multiple mutations (genetic markers) that are experimentally introduced into natural populations by small scale release experiments.

Tracking new genetic markers and the lineages harboring them will provide a clear, complete, and nearly "assumption free" information to address the spread of a new gene over time and space in natural settings. Among population genetics approaches, direct tracking of genes under natural conditions has unparallel power to resolve alternative hypotheses, but its technical demands throughout its development and application as well as its ethical implications cannot be justified in every case. Here we outline the basic components, prominent advantages, and main challenges of this approach because, in our view, natural vector populations perturbation studies are indispensable for the development of every genetic control strategy, and will have to be implemented prior to the introduction of a functional gene(s) to alter the vectorial capacity phenotype of the vector.

The development phase of the experimental release of new makers (thereafter, ERNM) involves (a) colonization of mosquitoes from the region where experimental release is planned, (b) inducing multiple mutations spread throughout the genome by low intensity irradiation or chemical mutagenesis (or by inserting stable genetic tags using molecular methods) across the genome of a number of specimens, (c) derive a few iso-female lines from specimens carrying induced mutations by inbreeding over ca. twenty generations (desirable range) to produce practically homozygous lines and insure removal of most severely deleterious mutations, (d) after the lines have been inbred for several generations (i.e., successful breeding for over ca. 7 generations in outbred organisms would ensure overcoming the inbreeding depression that causes small colonies to crash), a few dozens of the newly induced mutations are identified and (e) molecular assays are developed for genotyping of field collected specimens. Efficient genome scanning tools (e.g., DNA chips) will allow identifying and later monitoring dozens or even hundreds of these genetic markers, thus maximizing the number of "loci" and minimizing the number of mosquitoes to be released and analyzed. The derived lines are ready for experimental release in the region where they originated. The release may require only few hundreds of mosquitoes per line, so no mass production is required. The application phase involves (a) identifying three release sites ca. 60-100 km apart and coordinating the release with all the relevant parties, (b) removing the same (or larger) number of females to be released prior to the release date and releasing the set numbers from one to three lines in each release site, (c) large samples of adults will be taken periodically from every release site for genotyping to determine the markers frequencies, (d) adult sampling of nearby populations will follow findings showing that some of the new markers have reached set frequencies at the release sites. Monitoring will involve genotyping of mosquitoes collected by a flexible sampling scheme that increases in the area surrounding the release site based on the data from previous dates.

ERNM can provide direct information on contemporary gene flow of alleles with various selective values (expected to vary between neutral and mildly deleterious) across geographic distance and various putative barriers to gene flow such as that separating the molecular forms of *An. gambiae*. A central element in ERNM is the replication in three independent sites in the same region, that together with the change over generations, facilitates separating systematic change in allele frequency due to selection from stochastic change due to drift, hence, providing means to estimate the selective value of each marker (assuming similar marker's selective values and drift in the three sites). Thus, the effects of chromosomal location and the selective value of the marker on gene flow will be estimated. The data can also provide accurate estimates of the effective population size and the deme's geographical area, without being confounded by migration. The variation between populations in these parameters will be obtained. The experimental release will provide comprehensive and direct information on all key parameters required for prediction of the outcomes of different genetic control strategies. Apart from providing additional population genetics (e.g., recombination rates under natural conditions in relation to the chromosomal position and inversions) and ecological (e.g., dispersal, longevity) parameters, it will provide practical information on the behavior and viability of the released mosquitoes and the effectiveness of various release strategies. Finally, release experiments in West, Central, and East Africa will facilitate comparison of results from different geographical and ecological regions.

The value of the experimental release for genetic control programs cannot be overestimated despite its logistical demands (e.g., above), but it also involves ethical challenges that must be addressed. The most important is the possible increased risk of disease transmission and personal irritation due to (1) a larger number of mosquitoes in the released area, and (2) a higher threat associated with mutagenized mosquitoes. Unlike typical genetic control programs, the experimental release aims at a partial and temporary introduction of a fraction of the markers (mutations) into populations. Thus, a single release of up to several thousand mosquitoes is required. The overall number of females in the area will not increase since the number released will be matched by the same or larger number of females removed (prior to the release). Further, cumulative sampling for monitoring the change in markers frequencies definitely will reduce the number of vectors in the area. Unlike introducing a new functional gene with expected phenotypic effects, ERNM uses randomly "sprinkled" mutations induced by irradiation or chemical mutagenesis, or by inserting a stable marker into multiple sites throughout the genome. Such mutations are expected to consist primarily of deleterious, slightly deleterious, and neutral mutations and therefore present a safe material for release. Notably, released mosquitoes originated from an area within 100 km of the release site, thus the risk of introduction of adaptive genes into the release area is negligible.

Clearly, the possibility of introducing a beneficial mutation (for the mosquito) can not be ruled out, but we stress that it is a very remote possibility and making the mosquito a more dangerous disease vector is even more unlikely. However, this point needs to be further evaluated and weighed against the risk of every intervention. In the case of developing a genetic control strategy using a functional gene attached to a genetic drive mechanism, the benefit of ERNM appears to outweigh its risks. Finally there is the possibility to release males that carry new markers on the Y chromosome only, thereby "disconnecting" the marker from the female phenotype. While informative in its unique way, it will not address many of the issues addressed using markers spread throughout the genome. Nevertheless it can be a starting point. Although developed to meet the needs of a genetic control program, ERNM can revolutionize population genetic research, especially if it provides different results from those derived based on classical population genetics approach.

Overall Impact on Malaria Transmission Intensity and Disease Burden

The successful introduction of a transgene into *An. gambiae* across Africa does not imply removal of malaria from the top of public health priorities in the continent. In fact, the expected impact of a successful spread of a transgene on malaria transmission is not clear. Epidemiological models dating back to the classical model of Macdonald-Ross[85] have shown that considerable reduction in human exposure to infective mosquito bites is needed to achieve substantial impact on malaria morbidity and mortality in most parts of tropical Africa.[86,87] With this in mind and using a simple population genetical and epidemiological model, Boëte and Koella[24,25] demonstrated that even in conditions that allow the allele conferring refractoriness to reach fixation in the local vector population, the efficacy of refractoriness should be almost 100% (i.e., assuming no parasite escape from the refractory phenotype of its vector) for a significant effect on malaria prevalence.

Unlike classical means for vector control such as insecticide impregnated bednets or intra-domiciliary spraying that are directed to reduce exposure of people to infective bites by targeting anthropophilic and endophilic mosquitoes regardless of species, transgenesis-based methods target a single species. Hence, even if natural populations of *An. gambiae* became completely refractory to *Plasmodium* parasites Africa-wide (including all its chromosomal and molecular forms, and even extending this to the sibling *An. arabiensis* as well), other anophelines species will maintain transmission of malaria in large areas.[65,88] The importance of these 'neglected' vector species in contributing to the overall malaria transmission must be considered when the question of the benefits expected from the release of transgenic mosquitoes is discussed.

In addition to the members of the *An. gambiae* complex, at least three species are considered as vectors of epidemiological importance in Africa: *An. funestus*, *An. nili* and *An. moucheti*. In certain areas, these vector species may contribute more to disease transmission than the members of the *An. gambiae* species complex.[89-94] This is particularly the case in the humid savannas and forests of Central Africa, which remain largely unexplored.[95-97] One example of this situation that demonstrates how little we know on malaria vectors in Central Africa is the recent description, based on morphological and molecular evidences, of a new species, member of the *An. nili* group.[98] This newly described species appears to be the major malaria vector along rivers in South Cameroon. In such highly malaria endemic areas, eliminating malaria transmission by *An. gambiae* would change little the epidemiology of the disease and may even trigger unexpected worsening effects through insufficient decrease in transmission intensity.[99,100] Only if the transgenic approach proved successful in *An. gambiae* and is extended to the other vectors, then this strategy could realize its outmost impact on disease prevalence.

Finally, we point out that unlike conventional means of control such as insecticides, drugs or vaccines, we can do nothing to halt the spread of an undesirable effect brought about by the transgene spread in the vector populations. Designing a "recall mechanism", allowing halting the spread and possibly reversing it, would greatly improve the prospects and acceptability of the genetic control strategy.

Altogether, this discussion highlights serious limitations of our current ability to apply the genetic control strategy for malaria control in Africa. Current knowledge of vector populations and the epidemiology of malaria in Africa has lagged behind and its limitations call for caution when assessing the expected outcomes of a release of genetically altered vectors into the wild. However, the impressive progress in our understanding of the genetics and molecular biology of *Plasmodium falciparum*, its vectors and their interactions suggests that addressing these limitations is not beyond our reach.

References

1. Collins FH. Prospects for malaria control through the genetic manipulation of its vectors. Parasitol Today 1994; 10(10):3170-371.
2. Varmus H, Klausner R, Zerhouni E et al. Public health. Grand Challenge in Global Health. Science 2003; 302(5644):398-399.
3. James AA, Beerntsen BT, Capurro ML et al. Controlling malaria transmission with genetically-engineered, Plasmodium-resistant mosquitoes: Milestones in a model system. Parassitologia 1999; 41:461-471.
4. Coluzzi M, Costantini C. An alternative focus in strategic research on disease vectors: The potential of genetically modified nonbiting mosquitoes. Parassitologia 2002; 44:131-135.
5. Besansky NJ, Hill CA, Costantini C. No accounting for taste: Host preference in malaria vectors. Trends Parasitol 2004; 20(6):249-251.
6. Ribeiro JMC, Kidwell MG. Transposable elements as population drive mechanisms: Specification of critical parameters values. J Med Entomol 1994; 31:10-15.
7. Davis S, Bax N, Grewe P. Engineered underdominance allows efficient and economical introgression of traits into pest populations. J Theor Biol 2001; 212:83-98.
8. Sinkins SP, Godfray HC. Use of Wolbachia to drive nuclear transgenes through insect populations. Proc Biol Sci 2004; 271(1546):1421-1426.
9. Lanzaro G, Tripet F. Gene flow among populations of Anopheles gambiae: A critical review. In: Takken W, Scott T, eds. Ecological aspects for application of genetically modified mosquitoes. Wageningen Frontis Series. Vol. 2. Dodrecht, The Netherlands: Kluwer Academic Press, 2003:109-132.
10. Okanda FM, Dao A, Njiru BN et al. Behavioural determinants of gene flow in malaria vector populations: Anopheles gambiae males select large females as mates. Malar J 2002; 1:10.
11. Charlwood JD, Pinto J, Sousa CA et al. 'A mate or a meal' – Pregravid behaviour of female Anopheles gambiae from the islands of Sao Tomé and Principe, West Africa. Malar J 2003; 2:9.
12. Tripet F, Touré YT, Taylor CE et al. DNA analysis of transferred sperm reveals significant levels of gene flow between molecular forms of Anopheles gambiae. Mol Ecol 2001; 10:1725-1732.
13. Diabate A, Baldet T, Brengues C et al. Natural swarming behaviour in the molecular M form of Anopheles gambiae. Trans R Soc Trop Med Hyg 2003; 97:1-4.

14. Tripet F, Dolo G, Lanzaro GC. Multilevel analysis of genetic differentiation in Anopheles gambiae s.s. reveal patterns of gene flow important for malaria-fighting projects. Genetics 2005; 169:313-324.
15. Gould F, Schliekelman P. Population genetics of autocidal control and strain replacement. Annu Rev Entomol 2004; 49:193-217.
16. Coluzzi M, Sabatini A, della Torre A et al. A polytene chromosome analysis of the Anopheles gambiae species complex. Science 2002; 298:1415-1418.
17. Ayala FJ, Coluzzi M. Chromosome speciation: Humans, drosophila, and mosquitoes. Proc Natl Acad Sci USA 2005; 102(1):6535-6542.
18. della Torre A, Fanello C, Akogbeto M et al. Cytogenetic and molecular evidence of incipient speciation within Anopheles gambiae s.s. in West Africa. Insect Mol Biol 2001; 10(1):9-18.
19. della Torre A, Costantini C, Besansky NJ et al. Molecular and ecological aspects of incipient speciation within Anopheles gambiae: The glass is half full. Science 2002; 298:115-117.
20. della Torre A, Tu Z, Petrarca V. On the distribution and genetic differentiation of Anopheles gambiae s.s. molecular forms. Insect Biochem Mol Biol 2005; 35:755-769.
21. Krzywinski J, Besansky NJ. Molecular systematics of Anopheles: From subgenera to subpopulations. Annu Rev Entomol 2003; 48:111-139.
22. Powell JR, Petrarca V, della Torre A et al. Population structure, speciation, and introgression in the Anopheles gambiae complex. Parassitologia 1999; 41:101-114.
23. Tabachnick WJ. Reflections on the Anopheles gambiae genome sequence, transgenic mosquitoes and the prospect for controlling malaria and other vector borne diseases. J Med Entomol 2003; 40(5):597-606.
24. Boëte C, Koella JC. A theoretical approach to predicting the success of genetic manipulation of malaria mosquitoes in malaria control. Malaria Journal 2002; 1:3-9.
25. Boëte C, Koella JC. Evolutionary ideas about genetically manipulated mosquitoes and malaria control. Trends Parasitol 2003; 19(1):32-38.
26. Moreira LA, Wang J, Collins FH et al. Fitness of anopheline mosquitoes expressing transgenes that inhibit Plasmodium development. Genetics 2004; 166:1337-1341.
27. Boëte C. Malaria parasites in mosquitoes: Laboratory models, evolutionary temptation and the real world. Trends Parasitol 2005; 21(10):445-447.
28. Lambrechts L, Halbert J, Durand P et al. Host genotype by parasite genotype interactions underlying the resistance of anopheline mosquitoes to Plasmodium falciparum. Malar J 2005; 4:3.
29. Martinez-Torres D, Chandre F, Williamson MS et al. Molecular characterization of pyrethroid knock-down resistance (kdr) in the major malaria vector Anopheles gambiae s.s. Insect Mol Biol 1998; 7:179-184.
30. Chandre F, Manguin S, Brengues C et al. Current distribution of pyrethroid resistance gene (kdr) in Anopheles gambiae complex from West Africa and further evidence for reproductive isolation of the Mopti form. Parassitologia 1999; 41:319-322.
31. Fanello C, Petrarca V, della Torre A et al. The pyrethroid knock-down resistance gene in the Anopheles gambiae complex in Mali and further indication of incipient speciation within An. gambiae s.s. Insect Mol Biol 2003; 12:241-245.
32. Weill M, Chandre F, Brengues C et al. The kdr mutation occurs in the Mopti form of Anopheles gambiae s.s. through introgression. Insect Mol Biol 2000; 9:451-455.
33. Diabate A, Baldet T, Chandre F et al. Kdr mutation, a genetic marker to assess events of introgression between the molecular M and S forms of Anopheles gambiae (Diptera: Culicidae) in the tropical savannah area of West Africa. J Med Entomol 2003; 40(2):195-198.
34. Yawson AE, McCall PJ, Wilson MD et al. Species abundance and insecticide resistance of Anopheles gambiae in selected areas of Ghana and Burkina Faso. Med Vet Entomoi 2004; 18:372-377.
35. Diabate A, Baldet T, Chandre F et al. First report of a kdr mutation in Anopheles arabiensis from Burkina Faso, West Africa. J Am Mosq Control Assoc 2004; 20(2):195-196.
36. Meister GA, Grigliatti TA. Rapid spread of a P element/Adh gene construct through experimental populations of Drosophila melanogaster. Genome 1993; 36:1169-1175.
37. Brown EB, Bugeon L, Crisanti A et al. Stable and heritable gene silencing in the malaria vector Anopheles stephensi. Nucleic Acids Res 2003; 31(15):e85.
38. Curtis C, Coleman PG, Kelly DW et al. Advantages and limitations of transgenic vector control: Sterile males versus gene drivers. In: Boëte C, ed. Genetically modified mosquitoes for malaria control. Georgetown: Landes Bioscience, 2006; 6:60-78.
39. Catteruccia F, Nolan T, Loukeris TG et al. Stable germline transformation of the malaria mosquito Anopheles stephensi. Nature 2000; 405:959-962.
40. Catteruccia F, Godfray HCJ, Crisanti A. Impact of genetic manipulation on the fitness of Anopheles stephensi mosquitoes. Science 2003; 299:1225-1227.

41. Crow JF, Kimura M. Introduction to population genetics theory. New York: Harper and Row, 1970.
42. Waples RS. Genetic methods for estimating the effective population size of cetacean populations. In: Hoelzel AR, ed. Genetic ecology of whales and dolphins. Cambridge, UK: International Whaling Commission (Special Issue n°13), 1991:279-300.
43. Taylor CE, Touré YT, Coluzzi M et al. Effective population size and persistence of Anopheles arabiensis during the dry season in West Africa. Med Vet Entomol 1993; 7:351-357.
44. Nei M, Tajima F. Genetic drift and estimation of effective population size. Genetics 1981; 98:625-640.
45. Pollak E. A new method for estimating the effective population size from allele frequency changes. Genetics 1983; 104:531-548.
46. Caballero A. Developments in the prediction of effective population size. Heredity 1994; 73:657-679.
47. Berthier P, Beaumont MA, Cornuet JM et al. Likelihood-based estimation of the effective population size using temporal changes in allele frequencies: A genealogical approach. Genetics 2002; 160:741-51.
48. Taylor CE, Manoukis NC. Effective population size in relation to genetic modification of Anopheles gambiae sensu stricto. In: Takken W, Scott T, eds. Ecological aspects for application of genetically modified mosquitoes.Wageningen Frontis Series. Vol. 2. Dodrecht, The Netherlands: Kluwer Academic Press, 2003:133-146.
49. Tallmon DA, Luikart G, Beaumont MA. Comparative evaluation of a new effective population size estimator based on approximate bayesian computation. Genetics 2004; 167:977-88.
50. Waples RS. Genetic estimates of contemporary effective population size: To what time periods do the estimates apply? Mol Ecol 2005; 14(11):3335-3352.
51. Lehmann T, Hawley WA, Grebert H et al. The effective population size of Anopheles gambiae in Kenya: Implications for population structure. Mol Biol Evol 1998; 15(3):264-276.
52. Simard F, Lehmann T, Lemasson JJ et al. Persistence of Anopheles arabiensis during the severe dry season conditions in Senegal: An indirect approach using microsatellite loci. Insect Mol Biol 2000; 9(5):467-479.
53. Morlais I, Girod R, Hunt R et al. Population structure of Anopheles arabiensis in La Reunion island, Indian Ocean. Am J Trop Med Hyg 2005; 73(6):1077-1082.
54. Wondji C, Simard F, Lehmann T et al. Impact of insecticide treated bed nets implementation on the genetic structure of Anopheles arabiensis in an area of irrigated rice fields in the Sahelian region of Cameroon. Mol Ecol 2005; 14:3683-3693.
55. Rousset F. Genetic differentiation within and between two habitats. Genetics 1999; 151:397-407.
56. Nei M, Chesser RK. Estimation of fixation indices and gene diversities. Ann Hum Genet 1983; 47:253-259.
57. Donnelly MJ, Simard F, Lehmann T. Evolutionary studies of malaria vectors. Trends Parasitol 2002; 18:75-80.
58. Lehmann T, Hawley WA, Grebert H et al. The Rift valley complex as a barrier to gene flow for Anopheles gambiae in Kenya. J Hered 1999; 90(6):613-621.
59. Lehmann T, Licht M, Elissa N et al. Population structure of Anopheles gambiae in Africa. J Hered 2003; 94(2):133-147.
60. Wright S. Isolation by distance. Genetics 1943; 28:139-156.
61. Rousset F. Genetic differentiation and estimation of gene flow from F-statistics under isolation by distance. Genetics 1997; 145:1219-1228.
62. Rousset F. Genetic differentiation between individuals. J Evol Biol 2000; 13:58-62.
63. Slatkin M. Isolation by distance in equilibrium and nonequilibrium populations. Evolution 1993; 47:264-279.
64. Wang J, Whitlock MC. Estimating population size and migration rates from genetic samples over space and time. Genetics 2003; 163:429-446.
65. Fontenille D, Simard F. Unraveling complexities in human malaria transmission dynamics in Africa through a comprehensive knowledge of vector populations. Comp Immunol Microbiol Infectious Diseases 2004; 27(5):357-375.
66. Coluzzi M. The clay feet of the malaria giant and its African roots: Hypotheses and inferences about origin, spread and control of Plasmodium falciparum. Parassitologia 1999; 41:277-283.
67. Avise JC. Phylogeography: The history and formation of species. Cambridge, MA: Harvard University Press, 2000.
68. della Torre A, Merzagora L, Powell JR et al. Selective introgression of paracentric inversions between two sibling species of the Anopheles gambiae complex. Genetics 1997; 146:239-244.

69. Besansky NJ, Krzywinski J, Lehmann T et al. Semipermeable species boundaries between Anopheles gambiae and Anopheles arabiensis: Evidence from multilocus DNA sequence variation. Proc Natl Acad Sci USA 2003; 100(19):10818-10823.

70. Whitlock MC, McCauley DE. Indirect measures of gene flow and migration: Fst?1/(4Nm+1). Heredity 1999; 82:117-125.

71. Donnelly MJ, Licht MC, Lehmann T. Evidence for recent population expansion in the evolutionary history of the malaria vectors Anopheles arabiensis and Anopheles gambiae. Mol Biol Evol 2001; 18:1353-1364.

72. Wright S. Evolution and the genetics of populations: Variability within and among natural populations. Chicago: University of Chicago Press, 1978.

73. Balloux F, Lugon-Moulin N. The estimation of population differentiation with microsatellite markers. Mol Ecol 2002; 11:155-165.

74. Excoffier L. Special issue: Analytical methods in phylogeography and genetic structure. Mol Ecol 2004; 13:727.

75. Barton N. Genetic hitchhiking. Phil Trans R Soc Lond B Biol Sci 2000; 355:1553-1562.

76. Lanzaro GC, Touré YT, Carnahan J e al. Complexities in the genetic structure of Anopheles gambiae populations in West Africa as revealed by microsatellite DNA analysis. Proc Natl Acad Sci USA 1998; 95:14260-14285.

77. Wang R, Zheng L, Touré YT et al. When genetic distance matters: Measuring genetic differentiation at microsatellite loci in whole-genome scans of recent and incipient mosquito species. Proc Natl Acad Sci USA 2001; 98(19):10769-10774.

78. Gentile G, della Torre A, Maegga B et al. Genetic differentiation in the African malaria vector, Anopheles gambiae s.s., and the problem of taxonomic status. Genetics 2002; 161:1561-1578.

79. Costantini C, Sagnon NF, Ilboudo-Sanogo E et al. Chromosomal and bionomic heterogeneities suggest incipient speciation in Anopheles funestus from Burkina Faso. Parassitologia 1999; 41:595-611.

80. Cohuet A, Dia I, Simard F et al. Gene flow between chromosomal forms of the malaria vector Anopheles funestus in Cameroon, Central Africa, and its relevance in malaria fighting. Genetics 2005; 169:301-311.

81. Michel AP, Guelbeogo WM, Grushko O et al. Molecular differentiation between chromosomally defined incipient species of Anopheles funestus. Insect Mol Biol 2005; 14(4):375-387.

82. Wondji C, Simard F, Fontenille D. Evidence for genetic differentiation between the molecular forms M and S within the Forest chromosomal form of Anopheles gambiae in an area of sympatry. Insect Mol Biol 2002; 11(1):11-19.

83. Turner TL, Hahn MW, Nuzhdin SV. Genomic islands of speciation in Anopheles gambiae. PLoS Biol 2005; 3(9):e85.

84. Stump AD, Shoener JA, Costantini C et al. Sex-linked differentiation between incipient species of Anopheles gambiae. Genetics 2005; 169:1509-1519.

85. Macdonald G. The epidemiology and control of malaria. London: Oxford University Press, 1957.

86. Smith TA, Leuenberger R, Lengeler C. Child mortality and malaria transmission intensity in Africa. Trends Parasitol 2001; 17:145-149.

87. Trape JF, Pison G, Spiegel A et al. Combating malaria in Africa. Trends Parasitol 2002; 18:224-230.

88. Fontenille D, Lochouarn L. The complexity of the malaria vectorial system in Africa. Parassitologia 1999; 41:267-271.

89. Antonio-Nkondjio C, Awono-Ambene P, Toto JC et al. High malaria transmission intensity in a village close to Yaounde, the capital city of Cameroon. J Med Entomol 2002; 39:350-355.

90. Antonio-Nkondjio C, Simard F, Awono-Ambene P et al. Malaria vectors and urbanization in the equatorial forest region of south Cameroon. Trans R Soc Trop Med Hyg 2005; 99:347-354.

91. Carnevale P, Le Goff G, Toto JC et al. Anopheles nili as the main vector of human malaria in villages of southern Cameroon. Med Vet Entomol 1992; 6:135-138.

92. Cohuet A, Simard F, Wondji C et al. High malaria transmission intensity due to Anopheles funestus in a village of Savannah-Forest transition area in Cameroon. J Med Entomol 2004; 41(5):901-905.

93. Dia I, Diop T, Rakotoarivony I et al. Bionomics of Anopheles gambiae Giles, An. arabiensis Patton, An. funestus Giles and An. nili (Theobald) (Diptera: Culicidae) and transmission of Plasmodium falciparum in a Sudano-Guinean zone (Ngari, Senegal). J Med Entomol 2003; 40(3):279-283.

94. Mendis C, Jacobsen JL, Gamage-Mendis A et al. Anopheles arabiensis and Anopheles funestus are equally important vectors of malaria in Matola coastal suburb of Maputo, southern Mozambique. Med Vet Entomol 2000; 14:171-180.

95. Hay SI, Rogers DJ, Toomer JF et al. Annual Plasmodium falciparum entomological inoculation rates (EIR) across Africa: Literature survey, internet access and review. Trans R Soc Trop Med Hyg 2000; 94:113-127.

96. Coetzee M, Craig M, le Sueur D. Distribution of African malaria mosquitoes belonging to the Anopheles gambiae complex. Parasitology Today 2000; 16(2):74-77.
97. Levine RS, Townsend Peterson A, Benedict MQ. Geographic and ecologic distributions of the Anopheles gambiae complex predicted using genetic algorithm. Am J Trop Med Hyg 2004; 70(2):105-109.
98. Awono-Ambene P, Kengne P, Simard F et al. Description and bionomics of Anopheles (Cellia) ovengensis (Diptera: Culicidae), a new malaria vector species of the Anopheles nili group from South Cameroon. J Med Entomol 2004; 41(4):561-568.
99. Trape JF, Rogier C. Combating malaria morbidity and mortality by reducing transmission. Parasitol Today 1996; 12:236-240.
100. Reyburn H, Drakeley C. The epidemiological consequences of reducing the transmission intensity of P. falciparum. In: Boëte C, ed. Genetically modified mosquitoes for malaria control. Georgetown: Landes Bioscience, 2006; 8:89-102.

CHAPTER 6

Advantages and Limitations of Transgenic Vector Control:
Sterile Males versus Gene Drivers

Christopher Curtis,* Paul G. Coleman, David W. Kelly
and Diarmid H. Campbell-Lendrum

Abstract

Transgenesis might be used to produce fitter and more acceptable sterile males than those hitherto produced with radiation or chemosterilants. It is possible to engineer a dominant lethal construct which can be conditionally switched off so that males carrying it can be reared for release. Sterile males can eradicate pest populations provided that one can exclude immigrant, monogamous females that have already made a fertile mating outside the release area. "Urban island" populations of vectors may meet the required conditions for successful eradication. The genetic engineering of strains which are not susceptible to *Plasmodium* spp. development is also likely to be possible. For such 'refractory genes' to be useful it will be necessary to drive them to fixation so that they completely replace wild vector populations. A system for driving refractory genes through populations should require smaller releases to initiate the population replacement process than does the Sterile Male Technique (SIT), and the driving system should be "resistant" to the effects of immigration. Among the driving systems which have been suggested are: (i) negatively heterotic systems; (ii) uni-directional cytoplasmic incompatibility due to the bacterial endosymbiont *Wolbachia*; and (iii) transposons. An assumption underlying the driving of genes into populations is that the driver and the gene to be driven will remain genetically linked. In fact, some degree of recombination is inevitable and, if the driver without the refractory gene is fitter than the driver linked to this gene, the end result could be fixation of the driver alone, and loss of the refractory gene from the population with no reduction in disease transmission. We modelled the above three types of driving system with incomplete linkage to the refractory gene and with a fitness cost associated with that gene. We conclude that the systems will only confer permanent refractory protection if there is perfect linkage between the driver and refractory genes. There may be some public health benefits associated with a reduction in disease transmission as the refractory gene initially spreads through the vector population. However, within a time horizon of about 10 years, under a range of assumptions of fitness costs and recombination rates, our simulations show that any short-term gains associated with an increased frequency of the desired refractory genotypes are lost as the driving mechanism, freed from the costs associated with the transgene, drives itself to fixation and the refractory trait is lost from the population.

*Corresponding Author: Christopher Curtis—London School of Hygiene and Tropical Medicine, Keppel Street, London WC1E 7HT, U.K. Email: Chris.Curtis@lshtm.ac.uk

Genetically Modified Mosquitoes for Malaria Control, edited by Christophe Boëte.
©2006 Landes Bioscience.

Introduction

Proposed methods of genetic control of mosquito vector populations may aim either to:

a. suppress or eliminate the populations by large and repeated releases of males carrying dominant lethals which kill the progeny of matings to wild females (the Sterile Insect Technique, SIT); or:

b. render the population genetically harmless by release of limited numbers of mosquitoes carrying factors which prevent transmission of human pathogens by the females. This construct would be linked to a "genetic driving system" which would raise these desirable factors up to a frequency sufficiently high as to reduce R_0 below 1, and to counterbalance the effect of immigration of mosquitoes carrying the wild type genes which allow the pathogens to be transmitted.

Transgenic techniques should lead to (a) more effective and acceptable sterile males than those produced hitherto with radiation or chemosterilants, and (b) to dominant, monogenic factors which block pathogen development and could more feasibly be genetically linked to driving systems than the multigenic pathogen transmission blocking factors which have hitherto been selected from wild populations e.g., ref. 1.

Transgenic Sterile Males Based on Conditionally Repressible Dominant Lethality

Engineering Sterility

Sterility, as required for Sterile Insect Programs, has a very specific meaning. The sterile male must be able to compete to deliver sperm to wild females. The sperm and associated ejaculate must be fully functional, able to perform the whole range of activities that the spermatophore of nonsterile males perform, which might include: suppression off female sexual receptivity; competition with the sperm of other males for access to eggs; and, critically, fertilisation of the egg. Only after fertilisation should the 'sterility' of sterile males become apparent, as mutations carried by the sperm disrupt development before adulthood. Various means of introducing such dominant lethal mutations to the haploid sperm genome have been explored.

Gamma irradiation of pupae of screw worm flies has been highly successful in producing adult males which could compete adequately for wild mates, leading to the eradication of this very serious cattle pest all the way from Texas to Panama.[2] In mosquitoes, however, irradiation of pupae harms the competitiveness of the emerging adults.[3] Delaying irradiation until adulthood produces males which, at least in laboratory cages, could compete for mates.[4] However, putting millions of adult males through the irradiation process without damaging them would be technically and logistically challenging.

Chemosterilisation of pupae led to *Culex*,[5] *Aedes*[6] and *Anopheles*[7] males which could compete well for mates in the field. However, the alkylating agents used for chemosterilisation are, unsurprisingly given their mode of action, mutagenic. Though detectable residues in emerging adults are short lived[8] it has been claimed that they could still be biologically active.[9] Thus production of sterile males without the need to use such chemicals would seem to be more acceptable to regulatory authorities and the public.

Thomas et al[10] proposed the engineering of transgenic constructs which caused dominant lethality, but which could be de-activated by rearing the larvae in tetracycline. Such a construct has now been produced in *Aedes aegypti*.[11] It was found that when the males were mated to wild type females, the progeny survived in water without tetracycline until late in larval life, but at that time mortality was nearly 100%. Thus, after a release of the males, their progeny larvae would provide competition for the progeny larvae of wild type males. This contrasts with releases of conventional sterile males whose progeny die as early embryos, thus presumably reducing the density-dependent mortality of the progeny of the wild males and producing a "rebound" against the control effort. It is intended in the near future to test in cages the mating competitiveness of the transgenic strain with conditional dominant lethality.

An extra feature which can probably be added to this type of transgenic male is to make the lethality female specific. In that case one would remove the tetracycline from the larvae intended to produce batches for release, thus ensuring the elimination of biting females. Sexing systems already exist, based on sieving out the larger female pupae from the smaller males in culicines[12] or on translocation of an insecticide resistance gene on to the Y chromosome so that the relevant insecticide selectively kills female anophelines.[13] In production on the scale of hundreds of thousands or millions these systems have proved remarkably effective in *Culex*,[14] *Aedes*[15] and *Anopheles*.[16] These sexing systems were not perfect but, combining them with the above described female limited transgenic system, should allow an assurance that mass releases would not add biting females, even temporarily, to the wild population. In addition, such female limited lethality would allow survival of the heterozygous male progeny of released males and, in the next generation, half of their daughters would die, thus propagating some of the sterilising effect of a release for a few generations.

Contrary to a widespread belief, female monogamy (monandry) is not a requirement for the success of sterile insect technique, provided that sperm carrying dominant lethals are competitive with normal sperm. In fact, however, females of most Dipteran species show a strong tendency towards monandry and, for this reason, nonisolation of target populations of sterile male releases can prevent the achievement of high levels of egg sterility because immigrant females already mated to fertile males refuse remating after arrival in the area where large numbers of sterile males have been released (e.g., ref. 14). In the case of the Screw Worm Fly it was possible to create a "rolling front" of massive aerial releases so that most immigrants to areas just behind the "front" had themselves mated to sterile males. It is hard to believe that resources would ever be available to rear enough *Anopheles gambiae* to attack the vast problem of rural African malaria vectors on this basis. However, we see an important role for the sterile insect technique, especially now that the conditional dominant lethal constructs are becoming available, against relatively isolated but vectorially important mosquito populations. We think especially of urban mosquito populations carrying disease to large and ever-growing urban human populations, where the mosquito species in the urban area does not exist in the surrounding rural area, i.e., is an "urban island". Examples may exist in south India where *An. stephensi stephensi* is an important urban vector but hardly exists in rural areas[17] and an equivalent situation with *An. arabiensis* in southern Nigerian cities where the nearby rural vector populations are wholly *An.* gambiae.[18,19] *Ae. aegypti* populations in some urbanised areas in Asia and Latin America may be equivalent "urban islands" which could be eradicated without rapid reinfestation being likely.

Fitness Consequences of Engineered Sterility

The effectives of the sterile insect approach to suppressing the mosquito population will depend on the ability of the sterile males to compete with their wild counterparts for matings with wild females. As discussed above, the mating competitiveness, or fitness (although strictly speaking, sterile males have zero fitness even if they are highly competitive) may be compromised by the method of sterilization. Gamma radiation has significant deleterious effects in mosquitoes, and transgenic approaches offer the possibility of reducing (although not totally avoiding) these negative impacts on mating competitiveness. Mosquito transformation is mediated through the use of transposable elements (see section below) to integrate the engineered genetic construct that achieves sterility (or refractoriness to disease) into the mosquito genome. The potential fitness impact of transposon-mediated transgenesis can be divided into two components: (i) the physiological / toxicological burden of the construct itself; and (ii) the insertional mutagenesis effect of the transposition event.

The former will obviously be idiosyncratic to the particular construct and may indeed be intentionally high, as for sterile insect technologies such as RIDL.[10] Good design will clearly optimise the degree and schedule of physiological and toxicological effects of expression off the construct, and are not dealt with further in this section.

The fitness effects of insertional mutagenesis on the other hand, will vary not with the construct design, but by insertion site, and there is good evidence to suggest that they can be minimised or avoided at reasonable effort by creating sufficient different strains and then selecting the fittest.

Transposons tend to insert into transcriptionally active areas of the genome, where the chromosomal DNA is necessarily more accessible, and this raises the possibility that the insertion may have an effect through disruption of native gene function. At an extreme, the effects may be lethal. However, most insertions will have a much smaller effect on fitness, presumably because they disrupt nonessential genes, or because they insert nearby but not into the gene. Producing a healthy transgenic line therefore becomes a numbers game: how many independent insertional lines do we need to make before we get at least one that has suitable fitness?

Work done to study the effect of insertional mutagenesis in *Drosophila* gives us some insight into this question. In the best study of its kind to date, Lyman et al[20] measured the fitness costs of 706 independent single P-element insertion lines of *Drosophila melanogaster*. They used an isogenic base stock in which to generate the insertional lines (*Sam ry^{506}*), and the P-element transposon was marked with *ry$^+$*, which resulted in a detectable eye colour change to allow the P-element insertion to be tracked.

Each independent insertional line was grown up as heterozygotes or homozygotes in the presence of (i.e., in competition with) the wild type, and viability scored as the ratio of wild-type to transformed adults emerging. This measure is highly relevant to questions of productivity in culture for transgenic insects, but as a broad empirical summation of 'fitness', it seems likely to be a proxy for adult measures of fitness also, such as longevity and mating success.

To complicate matters somewhat, Lyman et al found that the *ry$^+$* marker appears to improve the viability of *Drosophila* (on average, the heterozygous lines were fitter than the wild-type). This means that the absolute effect of the transposon alone on viability cannot be measured relative to the wild-type control. However, if we argue that the insertional effects are largely recessive, then the viability of heterozygotes should approximate to the wild-type, and therefore heterozygote viability can be used as a proxy for the wild-type control.

The variance in fitness impact between heterozygotes and homozygotes certainly supports the argument that insertional effects are largely recessive: if not, then the variance would be expected to be similar for both homo- and heterozygotes; in fact it is an order of magnitude smaller (Fig. 1), and comparable with the wild-type. Another way of looking at this is to argue that, were the effects dominant, or at least codominant, then one should expect a correlation between the magnitude of the viability effect between heterozygotes and homozygotes of the same line. Again, no such relationship exists in the data, supporting the hypothesis that insertional effects are essentially recessive in character (estimate of the slope for chromosome 2 = 0.002 ± 0.02, $p = 0.9$; chromosome 3 = -0.01 ± 0.02, $p = 0.5$).

We can therefore tentatively proceed to estimate the insertional mutagenic effects of P-element insertions using the heterozygote viability as the control for the corresponding homozygote. This can be expressed as the percentage change in homozygote viability with respect to the heterozygotes (Fig. 2).

The distribution is clearly skewed towards zero reduction in viability, and extremely consistent between insertions on Chromosomes 2 and 3. The skewed distribution suggests two things. Firstly, that the frequency distribution appears to be bounded by zero; this makes biological sense—it is very easy to imagine how an alteration in native gene expression might be deleterious, but very difficult to imagine how it might be beneficial to broad measures of fitness such as competitiveness in culture. Secondly, the distribution suggests that if the production of 'healthy' transgenics is a numbers game, then the odds are stacked in our favour. Excluding those lines which are effectively homozygous lethal and which, in the normal course of mosquito transgenesis, would probably not be picked up in the screening

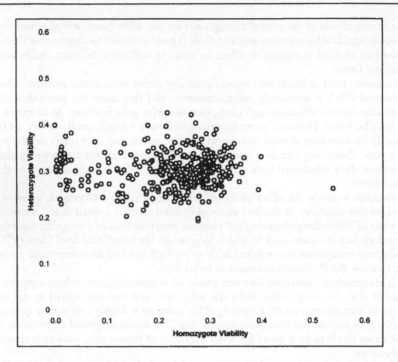

Figure 1. Reduction in viability of heterozygotes versus homozygotes of independent insertional lines, illustrating the order-of-magnitude difference in variance between homozygotes (20.4×10^{-3}) and heterozygotes (2.37×10^{-3}). Data from Lyman et al.[20]

in the first place, the median reduction in viability is just 15%. One third of independent homozygous transgenic lines isolated may have a fitness reduction of less than 10%, which is a fraction of the fitness costs estimated for the very small number of lines of transgenic mosquitoes currently studied.[21-24]

Another relevant piece of research was reported for transposon-mediated mutagenesis of the bacterium *Escherichia coli*.[25] A total of 226 mutants were independently derived from the 'wild-type' progenitor clone, and the fitness established in competition with the wild type in culture. Again, the distribution is highly skewed in favour of producing healthy transgenics, and apparently bounded by zero (Fig. 3). The study found no significant increase in fitness in any of the clones, but only a median reduction in fitness of 1%. While bacteria are clearly taxonomically distant from insects, the study has the merit that measures of fitness come from huge populations of bacteria (billions of individuals) and are therefore statistically much more powerful than the *Drosophila* study above. As a minimum, we can say that this study gives us some confidence that the median fitness reduction of 15% for the *Drosophila* study is not overly optimistic.

The effects of reduced mating competitiveness of engineered sterile males on the success of an SIT programme may be examined using simple mathematical modelling. Rogers and Randolph[26] provide a model of SIT against a pest population regulated by density dependent processes. In Figure 4, we adapt this model by assuming the release sterile males are less competitive than the wild type competitors. For a given released ratio (that is sterile males released per wild type male) increasing levels of impaired mating competitiveness result in decreasing effectiveness of the SIT programme. Transgenic approaches to population suppression offer the near-term prospect of highly competitive sterile male mosquitoes for use in effective SIT programmes.

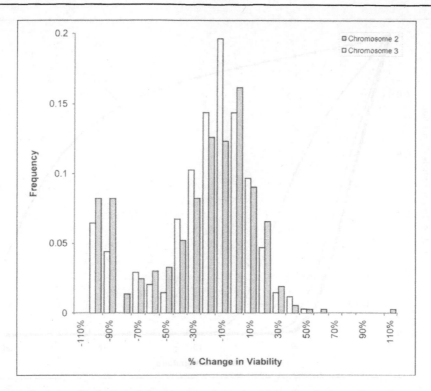

Figure 2. Frequency distribution of the percentage change in viability from heterozygotes to homozygote in 706 independent single *P*-element insertion lines of *Drosophila melanogaster* (365 on Chromosome 2; 341 on Chromosome 3). Data from Lyman et al.[20]

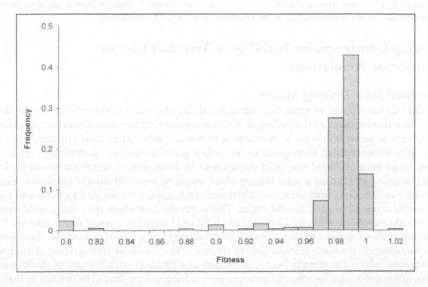

Figure 3. Frequency distribution of fitness effects of 266 Tn10 transposon-mediated mutagenesis on *Escherichia coli*. Data from Elena et al.[25]

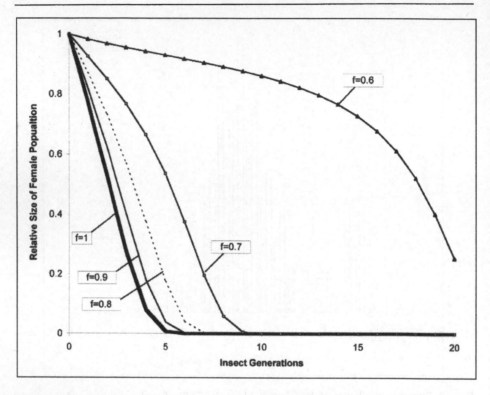

Figure 4. The importance of impaired relative mating competitiveness (f) of sterile males used in an SIT programme. The model used to simulate the effects of mass release of sterile males at a ratio of 3.5 per wild type male is that described by Rogers and Randolph,[26] in which the insect population is density dependent regulated. It can be seen, that as the relative mating competitiveness (f) changes from 1 (equal to the wild type) downwards, there is a decrease in the effectiveness of the SIT programme.

Driving Constructs for Inability to Transmit Disease into Vector Populations

The Need for a Driving System

The best known class of genes that reduce the ability of a vector to transmit disease are those that block development of the pathogen in the mosquito ("refractoriness"); other possibilities exist, such as genes which cause mosquitoes to bite animals rather than humans,[27] but for simplicity we refer to all such genes as 'refractory genes' hereafter. Theoretically, refractory genes could be introduced into wild populations by mass release, but there would be little point in doing so because a mass rearing plant would be more effectively employed rearing males for sterilisation; with these, an initial successful impact would lead to more and more favourable ratios of released to wild males. This is not the case where only the genetic nature, but not the numbers in the wild population, is being changed as a result of the releases. Furthermore, following mass release of mosquitoes carrying genes for refractoriness, there would be approximately the same vulnerability to reversal of the results by immigration of wild type mosquitoes, which has already been emphasised as a problem with the use of sterile males. Thus, linkage of a gene for refractoriness to a gene driving system should be the aim so that the desirable gene will spread from a limited "population seeding" release and the drive mechanism would tend to counteract the effect of immigration. Three mechanisms have been proposed for gene driving and are described below.

Negative Heterosis (= Underdominance for Fitness)

Serebovskii[28] in 1940 was the first to suggest a mechanism for positive selection upon a released genetic abnormality, using autosomal translocations with a viable and fertile homozygote but a semi-sterile heterozygote—a condition known as negative heterosis. Release of enough autosomal translocations, so that they were the majority type of chromosome, would be followed by selection for fixation of the translocation. Serebrovskii only viewed translocations as a means of population suppression by inherited partial sterility and he proposed to release just enough to approximate to the point of unstable equilibrium between translocation and wild type, so as maximise the time for which partial sterility would exist in the population. However, in 1968 Curtis[29] proposed to release enough translocations to deliberately tip the balance in their favour and to tightly link to the translocation (e.g., by an inversion) a refractoriness gene so that this desirable gene would be driven to fixation in the wild population.

More recently, Davis et al[30] have proposed a driving system based on a negatively heterotic transgenic system of two lethals (A and B which are unlinked constructs), with a suppressor of A being linked to lethal B and a suppressor of B being linked to lethal A. Thus, the pure strain would be viable because both lethals would be suppressed, but crosses to wild type would produce genotypes with an unsuppressed lethal, so that releases to raise the frequency of the transgenic strain above an equilibrium point should be followed by powerful selection for the released strain, which it is assumed would be equipped with refractoriness genes.

Bi-directional cytoplasmic incompatibility (i.e., sterility in both reciprocal crosses due to a maternally inherited factor) exists between many different geographical populations of *Culex pipiens* mosquitoes.[31] This is another example where a mixed population should show negative heterosis, with selection for the majority type. In 1970 Laven and Aslamkhan[32] proposed releasing an "integrated" strain which was both bi-directionally incompatible with the local wild strain and also carried a male-linked translocation complex which caused high sterility, but was not able to become homozygous because of its linkage to the male determining gene of *Culex*. Krishnamurthy and Laven[33] produced such a strain with cytoplasm of Paris origin, which was chosen for use against Indian wild populations. Curtis[34] showed, with an outdoor cage initially stocked with a wild type Indian strain, that sufficient releases of the integrated strain led to the fixation of this strain, despite its fitness load of >50% sterility due to the translocation complex. If there had been absolute cytoplasmic incompatibility, the translocation complex and the foreign cytoplasm of the integrated strain would have been effectively linked, but in fact occasional partial compatibility produced some "recombinant" individuals with foreign cytoplasm and no translocation. However, the repeated, relatively large, releases of the integrated strain made into the cage prevented a take-over by the fully fertile recombinant type and finally the cage population reached fixation for the integrated strain with its translocation—an approximate model of driving a pathogen blocking gene with some fitness load to fixation using a negatively heterotic driving system. The integrated strain was shown to have adequate competitiveness for mating in the field.[5] However, daily releases into two Indian villages over 3 months of 20,000-40,000 males, with about 40-80 females, only raised the frequency of egg rafts showing incompatibility and partial sterility due to the translocation to maxima of 62% and 8.6% respectively.[35] After consideration of various alternative hypotheses, the lack of the expected fixation of the integrated strain in the population was attributed mainly to immigration of already mated females from other villages.

Unidirectional Incompatibility and Wolbachia

A much more powerful driving force, which apparently could be initiated by minimal releases, is uni-directional cytoplasmic incompatibility, where matings of wild type females to released males are sterile but the reciprocal cross is fertile. The causative agent of this type of sterility was shown by Laven[31] to be maternally inherited. The strain whose females are not sterilised in a uni-directionally incompatible case would propagate its crossing type from all matings, in contrast to the type whose female is sterilised when crossed. Thus, the former type would be expected to have a selective advantage and to spread from a "seeding" release, and to sterilise the progeny of immigrants (unless the immigration rate is very high).[36] It should be noted that this type of selection is entirely different in nature from the frequency dependent selection which can arise from bi-directional incompatibility, as discussed in the previous section.

All wild *Culex pipiens* have been observed to be infected with *Wolbachia* symbionts. However Yen and Barr[37] were able to eliminate the symbionts by tetracycline treatment. They found that the resulting males became universally compatible with all other *Cx pipiens* but the females became compatible only with *Wolbachia*-freed males. Thus, for the reasons explained in the previous paragraph, a *Wolbachia* infected strain would be expected to be at a selective advantage to an uninfected one; indeed rapid spreading of a *Wolbachia* infected type has been observed in a *Drosophila* population.[38] Presumably the ability of *Wolbachia* to cause uni-directional incompatibility between *Wolbachia* infected and uninfected insects is a mechanism evolved by *Wolbachia* to favour its spreading throughout an insect population. In some mentions of *Wolbachia* as a possible driving system, it has been implied that it spreads by an infection process. However, this is not so—as far as is known, in nature, this symbiont is only vertically transmitted and spreading in populations is dependent entirely on selection due to uni-directional incompatibility and maternal inheritance. However, artificial horizontal transmission of *Wolbachia* has been possible by injection of eggs so as to reinfect an insect strain which had been made *Wolbachia*-free. After such artificial reinfection the expected compatibility properties were restored and horizontal transmission between genera (e.g., between *Aedes* and *Drosophila*) has been achieved.[39]

A survey of many wild and laboratory strains of *Anopheles* found none to be infected with *Wolbachia*[40] unlike many other groups of insects. Thus, if *Wolbachia* could be artificially introduced into an *Anopheles* strain and this were released, spreading of the infected state would be expected, due to the above described selection process. Attempts to inject *Wolbachia* into *Anopheles* eggs and thus to set up a sustained infection have so far not been successful, but S.Sinkins (personal communication) is optimistic that this could eventually be done.

In contrast to the situation with bi-directional cytoplasmic incompatibility, a nuclear gene (e.g., one which causes refractoriness to *Plasmodium*) and *Wolbachia* infection of the cytoplasm would have no tendency to stay linked together after release into a population—every fertile mating of wild type *Wolbachia*-free males to females of the *Wolbachia*-infected strain would produce "recombination" of wild type genes with *Wolbachia* infected cytoplasm. It may be possible to engineer refractoriness genes into the genome of *Wolbachia* or of mitochondria which are maternally inherited like *Wolbachia*. However, Sinkins and Godfray[41] consider it more feasible to engineer one of the determinants of cytoplasmic incompatibility so that it is placed on a nuclear chromosome closely linked to the nuclear gene which it is desired to drive into a population. Their idea is based on the above mentioned data of Yen and Barr[37] about the contrasting effects of the crossing properties of males and females after removal of *Wolbachia* from a mosquito strain. These results indicate that, in males, *Wolbachia* have the effect of making sperms unable to fertilise, unless the sperms are "rescued" by the action of a compatible type of *Wolbachia* in the female. It is not yet certain in what way the *Wolbachia* of different strains of mosquito vary so as to lead to the complex web of compatibility relationships which Laven[31] found between different *Cx pipiens* populations. However, the idea proposed by Sinkins and Godfray[41] is first to spread an artificially produced *Wolbachia* infected *Anopheles* strain through a wild vector *Anopheles* population and then to follow up with release of a strain with a chromosomally positioned factor which "rescues" sperms inactivated by the relevant *Wolbachia* strain The authors emphasise the "unreliability" of the process of transmission through the maternal cytoplasm and expect that a "rescue" factor on a chromosome would gain a selective advantage because of the absolute reliability of chromosomal inheritance. They predict that this selective advantage would drive the chromosome concerned to fixation and that a closely linked refractoriness factor would "hitch hike" to fixation also.

Transposable Elements

A transposable element (transposon) is located on a chromosome and tends to copy itself elsewhere on the chromosomes and hence to spread in populations. The *P* element is one such transposon which is known to have spread to fixation throughout the world's *Drosophila melanogaster* populations during the 20th century. It can be observed to spread to fixation in a cage population if initially mixed with non*P* carrying *Drosophila*.[42] *P* is not functional in *Anopheles* but other transposons are well known in these mosquitoes and have been proposed as driving systems for genes for refractoriness to *Plasmodium* genes.[43,44] A range of transposable elements, including *piggyback*, *minos*, *mariner* and *hermes*, have been successfully used to genetically transform mosquitoes. However, much remains to be discovered about these transposons and their suitability as effective drive mechanisms.[45] For example, transposition rate may be reduced by a repressor, as documented for the *P* element in *D. melanogaster* where mobility decreases after several generations due to the accumulation of a transposition inhibitor. Such effects would have important implications for the ability of a specific transposon to effectively drive refractoriness through a target population.

The Problems of Incomplete Linkage to Driving Systems and Reduced Fitness of Transgenes Which Cause Refractoriness

The Nature of the Problem

Genetic recombination between a driving system and a transgene which it is desired to drive into a population (or back mutation of the transgene to the wild type which can transmit the pathogen) would produce insects with the driving system alone. If the refractory gene had a fitness cost associated with it, the eventual result would be fixation of the driving system, but without the desired reduction in the vectorial capacity of the wild population.

To date, there has been very little investigation, whether experimentally or theoretically, into the limitations on these drive mechanisms. In particular, the possibility of separation of the driver mechanism and refractory genes has either not been considered, or it has been optimistically assumed that this could be prevented with an inversion.[29] In nature, separation could occur either through recombination in the case of transposons and negative heterotic systems, the incomplete inheritance of all cytoplasmic DNA in the case of *Wolbachia*, or inactivation of the refractory construct by random mutation in all three systems. Although such events are likely to be uncommon, they are inevitable in extremely large vector populations, given sufficient time. Once the link between the driving mechanism and the refractory gene is broken, the dynamics of gene driving will be different from those previously represented.

The fitness effects of transgene insertional mutagenesis has been discussed in an earlier section. In addition, expression of the refractory transgene might lead to a reduction in fitness. It is not clear whether it will be feasible to select a transgene causing 100% refractoriness to *Plasmodium* development in *Anopheles*, yet with negligible fitness cost. An *Anopheles* strain with the first transgene to be produced (a fluorescent marker) had severe reduction in fitness, but this was at least partly due to inbreeding depression, which may be avoidable with appropriate breeding schemes.[22] Of two transgenic strains with considerable reductions in susceptibility to rodent malaria,[23] one was reported to have normal fitness.[24] However, the transgene had been maintained in heterozygous condition by a selective breeding scheme and was then mixed with an equal number of wild types and, over the next five generations in a caged population, the frequency of the transgene did not decline significantly. With this breeding scheme few homozygotes for the transgene would have been produced and the test was primarily of heterozygous fitness of the transgene. Mutants commonly show reduced fitness only when made homozygous, and effective reduction in the vectorial capacity of a wild population would require that a refractoriness construct was driven to fixation of the homozygote. Further work is therefore needed to determine whether effective pathogen blocking transgenes with zero or minimal fitness reductions of the homozygotes can be engineered. Probably the prospects could be improved by arranging that the transgene is only switched on at the time of blood feeding, when *Plasmodium* gametocytes may be picked up and when they may succeed or fail to establish themselves as oocysts.

Mathematical Models to Test the Likely Effects of Recombination and Fitness Costs

Current models of genetic drive systems have tended not to include the effects of fitness costs associated with either the refractory transgene expression or the construct insertion effects. Rather, it has been suggested that reduced fitness due to the transgenes is of limited importance because a stronger or more efficient driving system could always be used to ensure the refractory trait is taken to fixation.[22] As we will show, this is not the case.

At present we only have hypothetical figures for fitness, but we consider it worthwhile to model the consequences of impaired fitness of transgenic lines for all the three drive systems. As with population suppression approaches (see Fig. 4), the inclusion of fitness costs reduces the effectiveness of the driver systems relative to the no-fitness-cost assumption. Fitness costs will raise the threshold conditions necessary to drive the refractory traits to fixation. This can be best shown for the negative heterosis, or underdominance system.

Davis et al[30] describe a number of possible configurations of the underdominance system, including the simplest system which is termed "extreme underdominance". In this system, if an engineered allele *A* was introduced into a wild population (all of gentotype *aa*) then the two homozygous genetoypes (*aa* and *AA*) are both viable, but the heterozgote (genetotype *Aa*) is not viable. Under the assumption of equal fitness of *AA* and *aa* genotypes, the predominant allele frequency will go to fixation (see Fig. 5A). In other words, if there were a single release of engineered *AA* individuals into a fully wild type population of *aa* genotypes, then the relative frequency of the *A* allele would have to be greater that 0.5 in order to drive to fixation. If we assume there is a fitness cost associated with the engineered *AA* genotype, then this threshold increases above 0.5. The critical threshold level to drive to fixation is defined by $1/[2 + (f - 1)]$, where f is the relative fitness of the *AA* genotype. The change in threshold allele frequency as a function of fitness cost is shown in Figure 5B. In the other underdominance systems described by Davis et al, such as the "nonhomologous" model, the inclusion of fitness costs results in the desired refractoriness trait being driven to a stable equilibrium which is below fixation. This is demonstrated in Figure 6A,B.

Next, we examine the importance of these fitness costs in a driving system with assumed low levels of recombination with the gene which it was attempting to drive to fixation. The first model, based on the Riberio and Kidwell[44] model of a transposon drive mechanism, describes changes in allele frequencies following an initial introduction of a defined proportion of individuals (i) which are homozygous for a single gene for refractoriness closely linked to a transposon (TR/TR), into a wild-type population in which all individuals are homozygous for the absence of both a transposon and the refractory gene (NS/NS). The populations are assumed to mate randomly, so that the genotype frequencies at the beginning of each subsequent generation are determined in accordance with Hardy-Weinberg expectations. We follow the previous modelling work[44] in specifying a simplification of the transposon copying mechanism, in which in all individuals heterozygous for the presence of a transposon (TR/NS), there is a probability (e) that the transposon will copy itself and the associated refractory or susceptible trait to the homologous chromosome, generating a disproportionate number of TR gametes.

To represent natural selection against refractory adults, the gene frequencies are adjusted by multiplying the frequencies of all individuals bearing the refractoriness gene by a relative fitness (f) between 0 and 1. The surviving individuals form gametes, whose genotype frequencies are

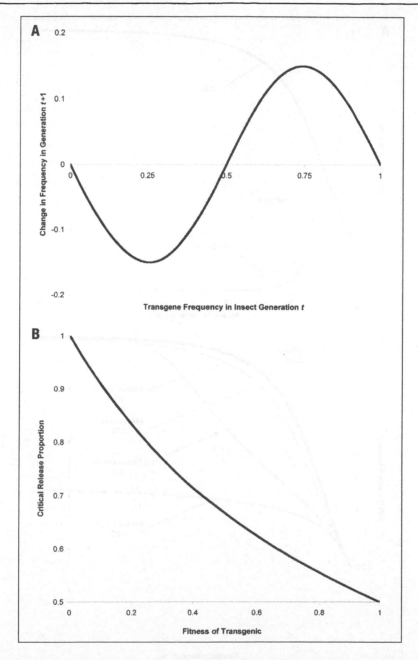

Figure 5. A) The existing model for simple underdominance has assumed that the genetic constructs and associated desirable genes are selectively neutral. Without fitness cost there are three fixed points for the driven allele, $A^* = 0, 0.5, 1$. The 0 and 1 points (extinction or fixation, respectively) are stable. The 0.5 point is unstable, so whatever is predominant, whether A or a, will go to fixation. The figure shows changes in refractory gene frequency in generation t+1 as a function of frequency in t. B) Importance of fitness costs in the extreme under-dominance model described by Davis et al.[30] The proportion of the population that must be transgenic for the trait to go to fixation is determined by the fitness of the transgenic genotype. With no fitness cost the proportion is >50%. With decreasing fitness, the proportion is increasingly greater than 50%.

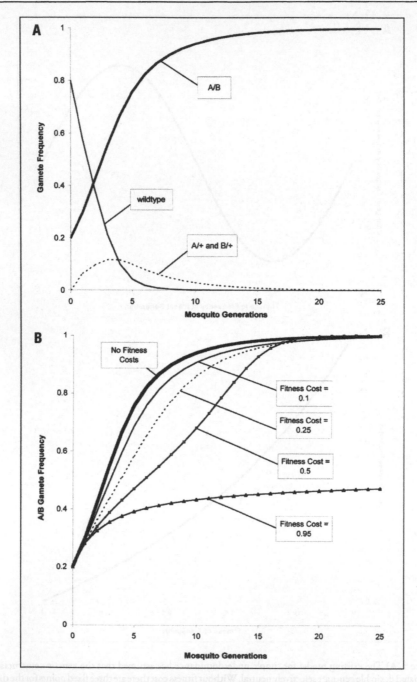

Figure 6. A) Example of nonhomologous underdominance with fitness costs which results in the refractory gene stabilising at an equilibrium level below complete fixation. Release frequency of AABB (engineered genotype) = 0.1, Additive fitness cost associated with dominance A = 0.3, B = 0.3, insertional effect assumed to be recessive gamma = 0.2, delta = 0.2. So genotype relative fitness aabb (wildtype) = 1, AaBb = 0.64, AaBB = 0.384, AABb = 0.384, AABB = 0.234. B) Relationship between fitness costs and equilibrium frequencies for the nonhomologous underdominance model.

determined by genotype frequencies in adults, adjusted by the probability of recombination (r) between the transposon and the refractory/susceptible gene. These gametes pair randomly to determine the initial genotypic frequencies of zygotes in the next generation. The model assumes complete density-dependence of population size, so that the number of individuals in each generation recovers to a constant level, which is independent of the relative frequency of its constituent genotypes.

We use the model to investigate two sets of parameter values: (1) a refractory trait conferring a low fitness penalty (relative fitness, f = 92.5%) but a relatively high recombination probability of 1 in 1000 per generation (i.e., r = 0.001), (2) a refractory trait conferring a higher fitness penalty (f = of 85%) but a much lower recombination probability of 1 in 10^6 (r = 0.000001). We set the rate of recombination at 0.1% in scenario 1 as we consider that in practice during laboratory testing of a new driver-transgene combination one could only hope to confidently exclude the existence of a recombination rate of about that order of magnitude.

Fitness costs are assumed to be multiplicative i.e., the relative fitness of homozygous refractory individuals is set as f^2. Under both fitness and recombination scenarios, the transposon drive mechanism ensures the refractoriness trait initially spreads despite the fitness cost imposed by the refractory trait (see Fig. 7). However, recombination eventually generates transposons in combination with susceptible wild-type gene that do not carry the refractory-associated fitness costs. Unshackled from the fitness constraint, these transposons drive themselves to fixation, replacing any refractory/transposon genotypes. The relative time taken for the refractoriness to initially spread before being replaced by the transposon/susceptible type is dependent on the initial release frequency (i), the fitness of the refractory mosquito (f) and the rate of recombination between the drive mechanism and the refractory loci (r).

Similar modelling approaches can be used to investigate gene driving using *Wolbachia*. The model is essentially the same as that described by Turelli et al,[46] but with separate estimation of the frequency of *Wolbachia* with and without refractoriness genes. We assume that a proportion (i) of individuals bearing dominant refractory genes linked to *Wolbachia* (WR) are introduced into a susceptible wild-type population without *Wolbachia* infections (NS). The model differs from that for transposon-driving in that the advantage of the *Wolbachia* is conferred through cytoplasmic incompatibility, whereby all matings of infected females with uninfected males result in fertile infected offspring, but the reciprocal cross is sterile. In the models shown here, we set the parameter for incompatibility as 1 (complete sterility). The possibility of the drive mechanism dissociating from the gene for refractoriness is determined not by recombination, but by the probability (p) of 'maternal disinheritance', where the endosymbiont is inherited without passing on the refractory gene—i.e., that WR gametes convert to WS. Another mechanism resulting in the conversion of WR to WS is random mutation resulting in inactivation of the refractory trait which will give WS alleles even if the refractory gene is encoded by engineering the *Wolbachia*. We consider a best-case version of the *Wolbachia* system by assuming no fitness costs associated with *Wolbachia* infection per se but rather with the inheritance of a refractory gene.

Figure 8 shows the temporal dynamics under the same scenario of fitness (i.e., f = 0.925 and f = 0.825) and effective recombination rates (p = 10^{-3} and 10^{-6}) as described for the transposon system. In contrast to a transposon driving-mechanism, the fitness of a *Wolbachia* driving mechanism depends partly on the frequency of *Wolbachia* within the population, and therefore the frequency of productive matings. The outcome is therefore relatively more sensitive to the initial release frequency of WR individuals. In Figure 8A,B, we assume an initial release of 10% and 30% WR individuals respectively. Again, as in the transposon model, the transient effect of initial refractory drive, followed by resurgence of susceptibility is seen as the driver dissociates from the costly cargo and goes to fixation in combination with the wild-type susceptible background.

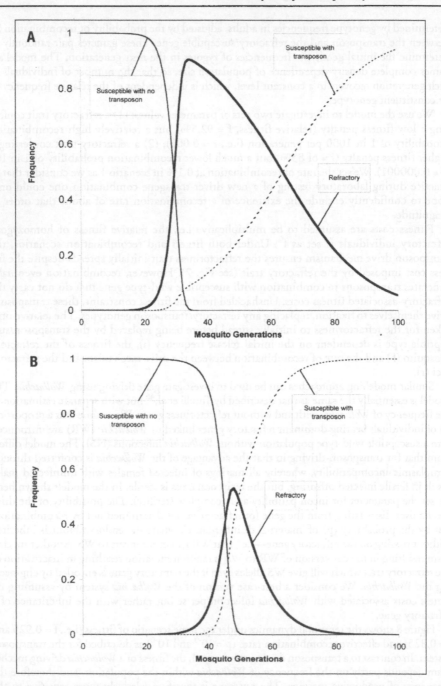

Figure 7. A) Predicted genotype frequencies over time, following an initial introduction of individuals (i = 10⁻⁶) homozygous for a transposon (T) with replication probability (e) = 0.8, initially linked to a refractory gene (R), with a fitness (f) of 0.925 relative to the wild type. Probability of recombination (r) is 1 in 1000 per generation. Mating is random and the population completely density dependent and stable. B) As above, but with a fitness (f) of 0.85 and a probability of recombination (r) of 1 in 10⁶ per generation.

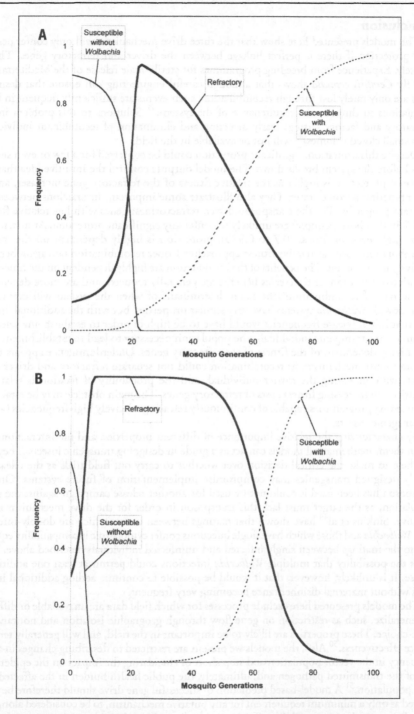

Figure 8. A) Changes in the *Wolbachia*/Refractory genotype with an initial release proportion of i = 0.1, relative fitness of the transgenic f = 0.925, and recombination rate of p = 0.001. B) As above, but with initial release of i = 0.3, fitness f = 0.825, and recombination rate p = 0.000001.

Conclusion

The models presented here show that the three drive mechanisms will only confer permanent protection if there is perfect linkage between the driver and refractory genes. This is unlikely. Experience from breeding programmes for sterile-male release of the Mediterranean fruit fly *Ceratitis capitata* shows that although careful engineering can ensure that desirable traits are only rarely lost through recombination, such events are sufficiently frequent in large populations to threaten the maintenance of the system.[47] Solutions to this problem in the laboratory and factory settings (early detection and elimination of recombinant individuals from small closed colonies)[48] will not be available in the field.

Despite this limitation, significant protection could be conferred for a few or even several years before the system breaks down. The model outputs confirm the intuitive idea that the degree of protection is higher as the relative fitness of the refractory gene increases, and as recombination becomes rarer. They also illustrate some important interactions between the different properties. For the transposon driver, refractoriness genes with low relative fitness (f <0.7 in the above example) are unlikely to confer any significant protection. At intermediate to high levels of fitness (0.7 to 0.975), protection is highly dependent on the rate of recombination. Only as relative fitness approaches 1 does recombination rate again become relatively unimportant. The benefits of the introduction are highly dependent on the time-scale considered. Over longer timescales benefits are generally reduced, and are more dependent on the rate of recombination (the main determinant of when the system will eventually break down). *Wolbachia* systems have very similar properties, but with the additional limitation that initial release frequencies would have to be high in order to establish any selective advantage for transgenic individuals: the population necessary to lead to establishment is in turn highly dependent of the fitness of the refractory genes. Underdominance appears to be a more robust mechanism, as recombination could not separate refractory and driver constructs without killing the carrier individual, and the probability of fixation is relatively insensitive to increasing fitness costs of refractory genes. The main obstacle may be establishing breeding programmes capable of continuously releasing relatively high frequencies (~3%) of transgenic insects.

By characterizing the relative importance of different properties, and the interactions between them, mathematical models can act as a guide in designing transgenic insects. They can also help to make the critical decision over whether to carry out field trials as the release of poorly designed transgenics may compromise implementation of future systems. Once a transposon has been used it could not be used for another release campaign against the same population, as the target must lack this transposon in order for the drive mechanism to be effective. Sinkins et al[49] have shown that matings between insects which are doubly-infected with *Wolbachia* and those which have single infections confer cytoplasmic incompatibility equivalent to the matings between singly infected and uninfected individuals described above. This raises the possibility that multiple *Wolbachia* infections could permit at least one additional release. It is unlikely, however, that it would be possible to continue adding additional infections without maternal disinheritance becoming very frequent.

The models presented here exclude processes for which field data are unavailable or difficult to generalize, such as restriction on gene flow through geographic isolation and nonrandom mate-choice. These properties are likely to be important in the field, and will generally tend to reduce effectiveness.[50] Also, the models we present are restricted to describing changes in allele frequency in the vector population and stop short of elucidating the impact on the epidemiology of the transmitted pathogen and, ultimately, the public health burden in the affected human population.[51] A model-based prediction of successful gene drive should therefore be considered as only a minimum requirement for any putative mechanism, to be considered alongside other factors, such as potential side-effects on other diseases and the wider environment, changes in the epidemiology of the target disease and cost-effectiveness relative to other disease-control measures.[51,52]

References

1. Collins FH, Sakai RK, Vernick KD et al. Genetic selection of a Plasmodium refractory strain of the malaria vector Anopheles gambiae. Science 1986; 234:607-612.
2. Wyss JH. Screw Worm eradication in the Americas - overview. In: Keng-Hong Tan, ed. Area-Wide Control of Fruit Flies and other Insect Pests. Penerbit Universiti Sains. Malaysia, Palang Pinang: International Atomic Energy Agency, 2000:79-86.
3. Patterson RS, Sharma VP, Singh KRP et al. Use of radiosterilized males to control indigenous populations of Culex pipiens quinquefasciatus: Laboratory and field studies. Mosq News 1975; 35:1-7.
4. Andreasen M, Curtis CF. Optimal life stage for radiation sterilization of Anopheles males and their fitness for release. Med Vet Ent 2005; 19:238-244.
5. Grover KK, Curtis CF, Sharma VP et al Competitiveness of chemosterilised and cytoplasmically incompatible-translocated (IS31B) males of Culex pipiens fatigans in the field. Bull Ent Res 1976a; 66:469-480.
6. Grover KK, Suguna SG, Uppal DK et al Field experiments on the competitiveness of males carrying genetic control systems for Aedes aegypti. Ent Exp Appl 1976b; 20:8-18.
7. Kaiser PE, Bailey DL, Lowe RE. Release strategy evaluation of sterile males of Anopheles albimanus with competitive mating. Mosq News 1981; 41:60-66.
8. LaBrecque G, Bowman WC, Patterson RS et al. Persistence of thiotepa and tepa in pupae and adults of Culex fatigans. Bull Wld Hlth Org 1972; 47:675-676.
9. Bracken GK, Dondale CD. Fertility and survival of Achaesanea tepidariorum on a diet of chemosterilised mosquitoes. Canad Ent 1972; 104:1709-1712.
10. Thomas D, Donnelly CA, Wood RJ et al. Insect population control using a dominant repressible lethal genetic system. Science 2000; 287:2474-2476.
11. Phuc HK, Andreasen M, Vass C et al. A dominant lethal system for mosquito control. Nature Genet, submitted.
12. Sharma VP, Patterson RS, Ford HR. A device for the rapid separation of male and female pupae. Bull Wld Hlth Org 1972; 47:429-432.
13. Seawright J, Kaiser PE, Dame DA et al. Genetic method for the preferential elimination of females of Anopheles albimanus. Science 1978; 200:1303-1304.
14. Yasuno M, MacDonald WW, Curtis CF et al. A control experiment with chemosterilised male Culex fatigans quinquefasciatus in a village near Delhi surrounded by a breeding free zone. Japan J Sanit Zool 1978; 29:325-343.
15. Ansari MA, Singh KRP, Brooks GD et al The development of procedures for mass rearing of Aedes aegypti. Ind J Med Res 1977; 65(suppl):91-99.
16. Dame DA, Lowe RE, Williamson DL. Assessment of released sterile Anopheles albimanus and Glossina morsitans. In: Pal R, Kitzmiller LB, Kanda T, eds. Proc XVI Internat. Cong Entomol Kyoto: Cytogeneics and Genetics of Vectors. New York: Elsevier Science Publ, 1981:231-248.
17. Ramachandra Rao. The Anophelines of India. Delhi: Malaria Research Centre, 1984.
18. Coluzzi M, Sabatini A, Petrarca V et al. Chromosomal differentiatin and adaptation to human environments in the Anopheles gambiae complex. Trans Roy Soc Trop Med Hyg 1977; 73:483-497.
19. Kristan M, Fleischman H, della Torre A et al. Pyrethroid resistance/susceptibility and differential urban/rural distribution of Anopheles arabiensis and An. gambiae malaria vectors in Nigeria and Ghana. Med Vet Ent 2003; 17:1-7.
20. Lyman RF, Lawrence F, Nuzhdin SV et al. Effects of single P-element insertions on bristle number and viability in Drosophila melanogaster. Genetics 1996; 143:277-92.
21. Irvin N, Hoddle MS, O'Brochta DA et al. Assessing fitness costs for transgenic Aedes aegypti expressing the GFP marker and transposase genes. Proc Natl Acad Sci 2004; 101:891-896.
22. Catteruccia F, Godfray HC, Crisanti C. Impact of genetic manipulation on the fitness of Anopheles stephensi mosquitoes. Science 2003; 299:1225-1227.
23. Ito J, Ghosh A, Moreira LA et al. Transgenic anopheline mosquitoes impaired in transmission of malaria parasites. Nature 2002; 417:452-453.
24. Moreira LA, Wang J, Collins FH et al. Fitness of anopheline mosquitoes expressing transgenes that inhibit Plasmodium development. Genetics 2004; 166:1327-1341.
25. Elena SF, Ekunwe L, Hajela N et al. Distribution of fitness effects caused by random insertion mutations in Escherichia coli. Genetica 1996; 102-103:349-58.
26. Rogers D, Randolph S. From a case study to a theoretical basis for tsetse control. Insect Sci Applic 1984; 5:419-23.
27. Pates HV, Takken W, Stuke K et al. Differential behaviour of Anopheles gambiae s.s. to human and cow odour. Bull Ent Res 2001; 91:289-296.

28. Serebrovskii AS. On the possibility of a new method for the control of insect pests. (In Russian). Zool Zh 1940; 19:618-630.

29. Curtis CF. Possible use of translocations to fix desirable genes in populations. Nature 1968; 218:368-369.

30. Davis S, Bax N, Grewe P. Engineered under-dominance allows efficient and economical introgression of traits into pest populations. J Theoret Biol 2001; 212:83-98.

31. Laven H. Speciation and evolution in Culex pipiens. In: Wright J, Pal R, ed. Genetics of Insect Vectors of Disease. Amsterdam: Elsevier, 1967:251-275.

32. Laven H, Aslamkhan M. Control of Culex pipiens pipiens and Culex p.fatigans with integrated genetical systems. Pak J Sci 1970; 22:303-312.

33. Krishnamurthy BS, Laven H. Development of cytoplasmically incompatible and integrated (translocated incompatible) strains of Culex pipiens fatigans for use in genetic control. J Genet 1976; 62:117-129.

34. Curtis CF. Population replacement in Culex fatigans by means of cytoplasmic incompatibility. 2. Field cage experiments with overlapping generations. Bull Wld Hlth Org 1976; 53:107-119.

35. Curtis CF, Brooks, Ansari MA et al. A field trial on control of Culex quinquefasciatus by releasae of males of a strain integrating cytoplasmic incompatibility and a translocation. Ent Exp and Appl 1982; 31:181-190.

36. Curtis CF, Sinkins S. Wolbachia as a possible means of driving genes into populations. Parasitology 1998; 116:S111-S115.

37. Yen JH, Barr AR The etiological agent of cytoplasmic incompatibility in Culex pipiens. J Invert Pathol 1973; 22:242-250.

38. Turelli M, Hoffman AA. Rapid spread of an inherited incompatibility factor in Californian Drosophila. Nature 1991; 353:440-442.

39. Sinkins S, Curtis CF, O'Neill SL. The potential applications of inherited symbiont systems to pest control. In: O'Neill SL, Hoffmann AA, Werren JH, eds. Influential Passengers. Oxford: Oxford Univ Press, 1997:155-175.

40. Sinkins SP. Wolbachia as a potential gene driving system. PhD thesis, Univ. London 1996.

41. Sinkins SP, Godfray HCJ. Use of Wolbachia to drive nuclear transgenes through insect populations. Proc R Soc Lond B 2004; 271:1421-1428.

42. Kidwell MG. Intraspecific hybrid sterility. In: Ashburner M, Carson HL, Thompson JN, eds. The Genetics and Biology of Drosophila, Vol 3. New York: Academic Press, 125.

43. Curtis CF, Graves PM. Methods for replacement of malaria vector populations. J Trop Med Hyg 1989; 91:43-48.

44. Ribeiro JM, Kidwell MG. Transposable elements as population drive mechanisms: Specification of critical parameter values. J Med Entomol 1994; 3:10-16.

45. Riehle MA, Srinivasan P, Moreira CK et al. Towards genetic manipulation of wild mosquito populations to combat malaria: Advances and challenges. J Exp Biol 2003; 206:3809-3816.

46. Turelli M, Hoffmann AA, McKechnie SW. Dynamics of cytoplasmic incompatibility and mtDNA variation in natural Drosophila simulans populations. Genetics 1992; 132:713-23.

47. Rössler Y. Recombination in males and females of the Mediterranean fruit fly (Diptera: Tephritidae) with and without chromosomal aberrations. Ann Entomol Soc Am 1982; 75:619-622.

48. Fisher K, Caceres C. A filter rearing system for mass reared medfly. In: Tan KH, ed. Area-Wide Control of Fruit Flies and Other Insect Pests. Penang, Malaysia: Penerbit Universiti Sains, 2000:543-550.

49. Sinkins SP, Braig HR, O'Neill SL. Wolbachia superinfections and the expression of cytoplasmic incompatibility. Proc Biol Sci 1995; 261:325-330.

50. Kiszewski AE, Spielman A. Spatially explicit model of transposon-based genetic drive mechanisms for displacing fluctuating populations of anopheline vector mosquitoes. J Med Entomol 1998; 35:584-590.

51. Boëte C, Koella JC. A theoretical approach to predicting the success of genetic manipulation of malaria mosquitoes in malaria control. Malar J 1998; 1:3.

52. Coleman PG, Alphey L. Genetic control of vector populations: An imminent prospect. Trop Med Int Health 2004; 9:433-437.

CHAPTER 7

Malaria-Refractoriness in Mosquito:
Just a Matter of Harbouring Genes?

Christophe Boëte*

Abstract

Amongst the last vector-based hopes for malaria control the most touted one is, without doubt, the use of transgenic mosquitoes able to kill/ block the development of malaria parasites and thus to interrupt malaria transmission. This potential solution is gaining some support because of advances in our comprehension of mosquito immunity and technical progress allowing mosquito modification. If the lack of a genetic drive is a technical problem that can easily be solved thus allowing an allele of interest to spread in natural populations, the difficulties inherent in this method, its possible consequences and its validity for malaria control remain questionable.

The technological aspects have been at the centre of the debate concerning the use of GM mosquitoes built to resist malaria infection but little research has been done until recently on the ecological and evolutionary aspects linked to this, still hypothetical method for malaria control. Indeed, let us consider that a mosquito enabled to be refractory to *Plasmodium falciparum* can be created in the laboratory, what then are the crucial factors for the successful field application of such a method? The spread of refractoriness, the evolutionary response of the parasite and obviously the effect on transmission and malaria epidemiology are certainly determinant. Up to now, the importance of these parameters has not been fully understood. However the evolutionary ecology of malaria-mosquito interactions and the knowledge concerning the mosquito's natural immune responses can bring some interesting insights into this debate. Obviously, one here may argue that the proposed solution relies mostly on artificial peptides, such as SM1[1] or the single chain antibody fragment N2scFv[2] and not so much on natural immune responses even if this is still evoked.[3] However if mosquitoes harbouring SM1 are able to reduce parasite oocyst number by about 80%, the only fully effective system described yet is the melanization response in selected lines of mosquitoes.[4] Moreover the ecological and evolutionary aspects of resistance against malaria parasites would certainly not be different whether it concerns artificial peptides or natural alleles. As proven, harbouring transgenes and malaria-resistant allele has been shown to be associated with a cost,[5-7] a classical view in evolutionary biology.

*Christophe Boëte—Institut de Recherche pour le Développement, Laboratoire Génétique et Evolution des Maladies Infectieuses, 911 avenue Agropolis, B.P. 64501 34394 Montpellier Cedex 05, France; Laboratory of Entomology, Wageningen University and Research Centre, P.O. Box 8031, Binnenhaven 7, 6700 EH Wageningen, The Netherlands; Joint Malaria Programme, Kilimanjaro Christian Medical Centre, P.O. Box 2228 Moshi, Tanzania. Email: cboete@gmail.com

Genetically Modified Mosquitoes for Malaria Control, edited by Christophe Boëte.
©2006 Landes Bioscience.

The Spread of Refractoriness

The first step for the success of genetically modified mosquitoes is the spread of refractoriness in natural populations of vectors. If research has focused on mechanisms conferring resistance against *Plasmodium* spp., little research has been done on the spread of an allele conferring refractoriness in natural populations of mosquitoes. Several methods have been considered for the release and maintenance of alleles of interest in mosquito populations. One of these is the use of a construction based on two-independent lethal alleles, each of them being linked to a suppressor allele of the other. If the two alleles are present, the individual that carries them survives and over a certain frequency within the population, the system can be maintained.[8,9] Linking an allele conferring refractoriness to such a system could permit its spread. More recently the use of endonucleases (Homing Endonucleases Genes) has been considered to favour the spread of allele of interest in vectors populations.[10] An alternative method could be the use of *Wolbachia*[11] but none so far have been found in *Anopheles*. Finally, most actual hopes depend on the use of a tandem, composed of a transposable element and an allele of refractoriness,[12] a solution for which mathematical models have been designed.[13-15] These models enable the determination of the conditions under which refractoriness can spread in mosquito populations. Thus, whereas the spread of an allele coding for refractoriness will depend on its associated costs and benefits,[15,16] when it is linked to a transposable element its fixation will depend mainly on the segregation bias. Also it appears relatively easy 'under ideal conditions' to invade a population with an allele conferring resistance to *P. falciparum*.[15,17] However the goal of a control program is not the spread of one or several genes of refractoriness against malaria in mosquito population but to reduce malaria burden in human populations.

So, what are the effects on the spread of such an allele on malaria epidemiology? The model developed by Boëte and Koella[15] gives us some insight on the effect on the epidemiology. Indeed the impact on prevalence in the areas of intermediate or high transmission is significant only if the efficacy is nearly total (Fig. 1). Also the major parameters affecting the outcome of a release of transgenic mosquitoes built to be resistant against malaria and enabled to invade the natural population are the efficacy of refractoriness and the initial level of transmission (i.e., the level of transmission when transgenic mosquitoes are introduced). This brings us back to the Ross-MacDonald equation of malaria transmission[18] which shows that this result is not surprising. Indeed, the basic reproductive number (R_0) and the human prevalence (y) are linked by a highly nonlinear relation (Box 1), therefore R_0 needs to decrease strongly in order to make any change in the human prevalence. Moreover, describing R_0 (Box 1) shows that both the efficacy of infection (b) and the number of susceptible mosquitoes (m) are linear terms. Modifying them (which would be done by the use of transgenic resistant mosquitoes) is also much less efficient than reducing the biting rate (a) or increasing the mosquito mortality (μ).

After the spread of refractoriness is achieved, the crucial factor is the efficacy of the mosquito immune response.

The Efficacy of Refractoriness

Until now, the best efficacy of refractoriness against *Plasmodium* spp. has been established with the natural immune response of encapsulation/melanization. Indeed a laboratory strain of *Anopheles gambiae* has been selected for the encapsulation/melanization of *P. cynomolgi* and it is also totally refractory to South American strains of *P. falciparum* and partially to African strains.[4]

However, will the genes that confer malaria-refractoriness (at least to some strains of parasites) function independently of environmental conditions?

Environmental Effects

As is the case for many invertebrates,[19] environmental conditions influence the mosquito immune response. Any stress during the larval stage[20] leads to a decrease in the mosquito's ability to melanize a foreign body. Thus, in mosquito lines selected for refractoriness,[4] a reduction in food availability from 100% to 25% of a standard diet reduces the proportion of mosquito

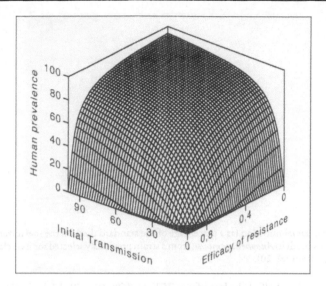

Figure 1. Effect of the fixation of an allele coding for malaria-refractoriness in a mosquito population on the human prevalence. If the allele has invaded the mosquito population, it has a significant impact on human prevalence only if the efficacy of refractoriness is close to 1 all the more since the initial transmission (the one observed in the absence of resistance from the mosquito) is high. (See ref. 15 for numeric values of the parameters.)

Box 1. The epidemiological aspects related to the spread of an allele conferring 'malaria-refractoriness' to mosquito

The intensity of transmission, R_0, and the human prevalence are linked via a non-linear relation (Equ. 1), thus any decrease in human prevalence (y) requires, especially in endemic areas, a dramatic decrease in the level of transmission.

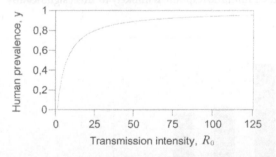

$$y = \frac{R_0 - 1}{R_0 + \dfrac{a}{\mu}} \qquad (1)$$

Moreover, to use genetically modified mosquitoes able to resist a malaria infection is equivalent to reduce the efficacy of the infection from human to mosquitoes (b) or the number of mosquito per human (m). Both those parameters are linear in the definition of the basic reproductive number (Equ. 2) and their modification is less efficient that an intervention aiming at increasing mosquito mortality (μ) or decreasing the biting rate (a).

$$R_0 = \frac{ma^2 b e^{-\mu T}}{r\mu} \qquad (2)$$

(T) and (r) are respectively the developmental time of the parasite and the recovery rate of human.

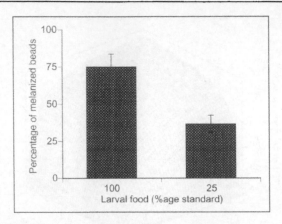

Figure 2. Effect of larval nutrition (as a percentage of the standard diet) on the melanization response of CM-25 Sephadex beads in *Anopheles gambiae*, from a strain previously selected for its melanization ability (figure modified from ref. 20).

melanizing more than half of the bead from 75% to 36% (Fig. 2). Melanization ability is also reduced when the adult mosquito is confronted with alimentary[21] or environmental[20] stress. Thus, it appears fundamental to measure the norms of reaction of the potential allele of interest[22] to determine the range of phenotypes that this allele could provide. Such a work has been done by *Drosophila* geneticists[23] but such information is lacking concerning the vectorial competence of the recently created transgenic mosquitoes.

Demographic Effects

Furthermore, mosquito's immune response weakens dramatically with age.[21,24] Indeed in about five days, the melanization ability can decrease by more than 70% in a malaria-susceptible and also in a refractory line (Fig. 3). Thus wild blood-fed females[25] show a decrease of 25% in their ability to melanize a foreign body between emergence and the age of four days. As most mosquitoes will become infected when they are already more than 5 days old, even in an area of high transmission,[26] such a decrease in their immune response is unlikely to affect significantly the malaria epidemiology.

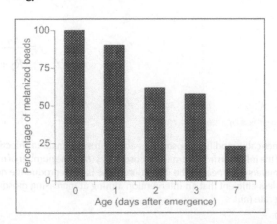

Figure 3. Effect of age after emergence on the melanization response of CM-25 Sephadex beads in *Anopheles gambiae*, from a strain previously selected for its melanization ability (figure modified from ref. 21).

It also appears that if technically it may be possible to favour the spread of an allele of interest in a natural population of mosquitoes, its efficacy may not be total and the reduction of prevalence very low. Therefore, engineering a high (at least lab-based) efficiency of resistance is not the only headache.

Efficacy and Number of Species

If the WHO has planned the use of GM mosquitoes for malaria control in a few years, it remains limited to *Anopheles gambiae s.s.* This choice reflects the importance of this sub-species in malaria transmission in Africa, but in fact dozens of species are vectors of malaria in the world. To make matters worse, in most areas, several species are responsible for malaria transmission. It has indeed been shown in Tanzania that up to 4 species can contribute to transmitting malaria in one site.[27] To render a species refractory to malaria parasite in such a setting will probably affect very lightly (or not) transmission. Indeed, if we consider that 2 vectors are equally responsible for transmission in a given area, to have one completely refractory is equivalent to having a total efficacy of 50% in the whole population. As seen previously, the efficacy of refractoriness has to be nearly total to have a significant effect on human prevalence.[15] Thus, in numerous situations, there is not one but several species of mosquitoes that need to be transformed to generate a significant decrease of the malaria burden.

Stability of the System

Finally, a critical point with this method using a transposable element and an allele conferring refractoriness is the stability of such a system. Indeed it is nearly impossible to assure the link between the allele of interest and the system used for its spread. It is very likely that recombination will produce individuals where the link between the allele and its driver will be broken. This should therefore be followed by the spread of the transposable element alone, making the construct impossible to reuse.

What about the Parasite?

Are All Parasites the Same?

Up to now the best efficiency against *Plasmodium* has been obtained in the laboratory with a strain of *Anopheles gambiae* selected to encapsulate *P. cynomolgi*.[4] This strain is also refractory to South American strains of *P. falciparum* but only partially to African strains of the parasites (Fig. 4). This intra-specific variability has been interpreted as a lack of recognition of the parasites strains by the host. This result also indicates the possibility of genotype by genotype interactions that have also been suggested in a study showing that different QTLs are involved in the mosquito's response against different species of malaria parasite.[28] Finally a recent paper has corroborated the existence of such genotype by genotype interaction between *P. falciparum* and *Anopheles gambiae*.[29]

If such relationships occur when using transgenic mosquitoes, the creation of different mosquito strains to cover the variants of *P. falciparum* should be envisaged. This technique would certainly then lead to an arms race between the two partners of the interaction.[30] Thus, in order to maintain effective parasite control, new systems of refractoriness should then be created. Moreover, as in the case of vaccine use,[31] such a programme should be completed with a survey of the parasites population(s) in order to determine any changes in the efficacy of refractoriness and to track any changes in parasite virulence that could arise from this system. Finally nothing is known on the effects of this introduced resistance against *P. falciparum* on the other human plasmodial species and on the modifications it may lead on the interactions between *P. falciparum* and the other 'minor' plasmodial species.[32a]

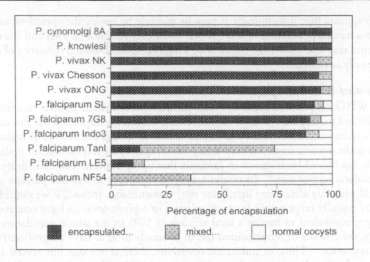

Figure 4. Lack of recognition or differential ability to suppress mosquito's melanization response? Measurement of the refractoriness level of an *Anopheles gambiae* strain selected for refractoriness against a strain of *P. cynomolgi* (first bar) against several plasmodial species and strains of *P. falciparum* from the Old and the New World (modified from ref. 4).

The Evolutionary Response of the Parasite

Apart from the question of the efficacy of refractoriness in natural populations of mosquitoes, one may wonder if, facing an increased resistance the parasite may respond by a suppression or escape mechanism. Indeed immunosuppression exists in many parasites that affect humans[32b] such as HIV,[33] *P. falciparum*[34,35] or the etiological agent of lepra *Mycobacterium leprae*.[36] In insects, the best examples of immunosuppression can be found in dipteran or hymenopteran parasitoid species, whose larvae develop within their insect host.[37] Indeed, when laying their eggs within their host, parasitoids inject poly-DNA virus or proteins[38,39] that leads to the destruction of cells involved in the immune response.[40-42] A similar phenomenon could explain the quasi absence of melanized ookinetes or oocysts of *P. falciparum* in *Anopheles gambiae* mosquitoes that are nevertheless able to melanize inert negatively charged CM-25 Sephadex beads.[25] A hypothetical reduction or suppression of melanization has already been suspected[43] and has been shown to occur in the model system *Aedes aegypti* and *P. gallinaceum* by two complementary experiments.[24,44] They have revealed that melanization response was suppressed by a direct action of ookinetes or young oocysts of *P. gallinaceum* and also by blood factors (Figs. 5, 6). This mechanism of immunosuppression is both stage-specific and systemic as the melanization response was tested in the thorax whereas the parasite is found in the abdomen. Such a method of escaping the immune response may also occur with any introduced artificial refractoriness.

Another concern is the fact that mounting a strong immune response against the malaria parasite may have detrimental effects. Indeed our experiment[44] revealed that mosquito mortality was significantly higher in the group where melanization was the strongest. This may be explained by the fact that cytotoxic molecules are released during the encapsulation response that have negative effects in an open-circulatory system.[45] Such a cost has already been observed in bumblebees, where in undernourished individuals, inducing an immune response leads to a reduction in survival.[46] This result suggests that the parasite, as a result of the immunosuppression induced, increases its own survival and also that of its vector, by reducing its mortality associated with the immune reaction. Thus, the parasite compromises the evolutionary response of resistance by the host against the evasion of the parasite.

It therefore appears interesting to reexamine the results of the selection experiment for refractoriness against *P. cynomolgi* by Collins et al.[4] Indeed the melanization response is species

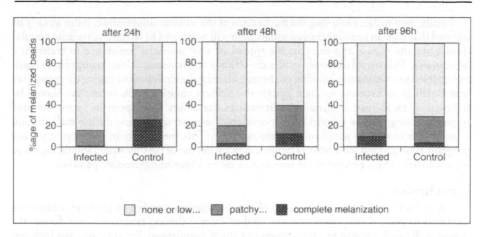

Figure 5. Immunoreduction in *Aedes aegypti* by *P. gallinaceum*. Each panel represents the melanization response of CM-25 Sephadex beads in mosquitoes at different time after an infectious or uninfectious (control) blood-meal. The melanization ability is determined according to the degree of melanization of the bead (none or low, patchy and complete). It appears that the infected blood-meal is drastically reducing the mosquito melanization response when the parasite is at the ookinete or young oocyst stage. (Figure modified from ref. 24.)

and strain-dependant and if it can be seen as genotype by genotype interactions, one can also express the hypothesis of a different ability to suppress the mosquito melanization response in the different strains of malaria parasites as seen above. Obviously it remains to be demonstrated that other species of *Plasmodium* spp. and especially *P. falciparum* can suppress their mosquito vector's melanization response and that this immunosuppression has a genetic basis.

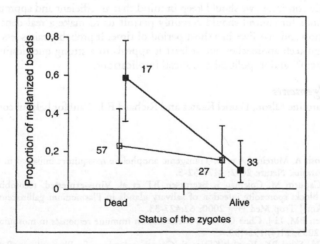

Figure 6. Direct and indirect immunosuppression by *P. gallinaceum* in *Aedes aegypti*. The black dots represent the mosquitoes having blood-fed on a healthy plasma whereas the open dots represent the ones having blood fed on an infected one. The melanization response is measured 24h after the blood meal. Both the infected plasma and the parasite (at the ookinete or young oocyste stage) lead to a decrease in the mosquito ability to melanize CM-25 Sephadex bead injected in the mosquito's thorax. This highlights a stage-specific, systemic, direct and indirect immunosuppression. The numbers of mosquitoes are given near each dot (figure modified from ref. 44).

Finally even without knowing the mechanism of the immunosuppression, it can already be assessed that such an immune evasion mechanism will occur and be selected in parasite populations after the release and the possible spread of any 'artificial' resistance in transgenic mosquitoes. This would obviously reduce the efficacy of the system. Using transgenic mosquitoes enabled to resist malaria infection can be considered to be equivalent to forcing the mosquito's investment in an immune response against the malaria parasite. Such an increase should be followed by an increase in parasite investment in the ability to suppress its vector immune response, as suggested by a model initially developed to better understand the low melanization of ookinetes in natural population of mosquitoes.[47] This means that in the long term, any program aiming at reducing malaria transmission by the use of GM mosquito should expect to see a reduced success because of the selection of more 'immunosuppressive' parasites.

Conclusion

Thus, from a purely technical point of view, it appears possible to generate mosquitoes modified to resist *P. falciparum* infection and to carry a driving system that will favour the spread of the allele conferring refractoriness in natural populations. From a molecular biologist point of view, this gives optimism for malaria control, but probably not for a population biologist and for an epidemiologist. Indeed, whether transgenic mosquitoes can be efficiently deployed in the field, i.e., lead to a significant and sustainable change in malaria transmission, remains unclear.[48] Once the allele conferring refractoriness has spread in a population of mosquitoes, the major question remains its efficacy in any environmental conditions, at any age of the mosquito, against all genotypes of parasites and its ability to cope with a potential evolutionary response by the parasite.

These results highlights the fact that a better understanding of the evolutionary ecology of mosquito refractoriness could allow us to determine if transgenic mosquitoes are a useful way of controlling malaria. It should also permit us to determine if continuing efforts in the creation of GM mosquitoes for malaria control is worthwhile.

Finally the use of transgenic mosquitoes for malaria gives rise to many questions and little hope for a change in malaria situation in the coming years. If the search for new methods to control malaria continues, we should keep in mind that an efficient and appropriate use of the existing methods for control should certainly permit us to make a real dent in the current malaria situation and save lives in a short period of time. It probably raises less scientific questions than high-tech approaches but at least it appeals to a strong questioning of the global economic system[49] and its political and social implications.

Acknowledgements

I thank Caroline Ellson, Daniel Keates and Richard E.L. Paul for helpful comments on this manuscript.

References

1. Ito J, Ghosh A, Moreira LA et al. Transgenic anopheline mosquitoes impaired in transmission of a malaria parasite. Nature 2002; 417:452-5.
2. de Lara Capurro M, Coleman J, Beerntsen BT et al. Virus-expressed, recombinant single-chain antibody blocks sporozoite infection of salivary glands in Plasmodium gallinaceum-infected Aedes aegypti. Am J Trop Med Hyg 2000; 62:427-33.
3. Christensen BM, Li J, Chen CC et al. Melanization immune responses in mosquito vectors. Trends Parasitol 2005; 21(4):192-199.
4. Collins FH, Sakai RK, Vernick KD et al. Genetic selection of a Plasmodium- Refractory Strain of the malaria vector Anopheles gambiae. Science 1986; 234:10607-610.
5. Catteruccia F, Godfray HCJ, Crisanti A. Impact of genetic manipulation on the fitness of anopheles stephensi mosquitoes. Science 2003; 299:1225-1227.
6. Moreira LA, Wang J, Collins FH et al. Fitness of anopheline mosquitoes expressing transgenes that inhibit Plasmodium development. Genetics 2004:1337-1341.
7. Irvin N, Hoddle MS, O'Brochta DA et al. Assessing fitness costs for transgenic Aedes aegypti expressing the GFP marker and transposases genes. Proc Natl Acad Sci USA 2004; 101.

8. Curtis CF. Possible use of translocations to fix desirable genes in insect pest populations. Nature 1968; 218:368-9.
9. Davis S, Bax N, Grewe P. Engineered underdominance allows efficient and economical introgression of traits into pest populations. J Theor Biol 2001; 212:83-98.
10. Burt A. Site-specific selfish genes as tools for the control and genetic engineering of natural populations. Proc R Soc Lond B Biol Sci 2003; 270:921-928.
11. Curtis CF, Sinkins SP. Wolbachia as a possible means of driving genes into populations. Parasitology 1998; 116:S111-5.
12. Curtis CF. The case for malaria control by genetic manipulation of its vectors. Parasitology Today 1994; 10:317-373.
13. Ribeiro JM, Kidwell MG. Transposable elements as population drive mechanisms: Specification of critical parameter values. J Med Entomol 1994; 31:10-6.
14. Spielman A. Release rations employed for genetically modifying populations of mosquitoes. In: Takken W, Scott TW, eds. Ecological aspects for application of genetically modified mosquitoes. Dordrecht: Kluwer Academic Publishers, 2003:9-12.
15. Boëte C, Koella JC. A theoretical approach to predicting the success of genetic manipulation of malaria mosquitoes in malaria control. Malaria J 2002; 1:3.
16. Boëte C, Koella JC. Evolutionary ideas about genetically manipulated mosquitoes and malaria control. Trends Parasitol 2003; 19:32-38.
17. Hahn MW, Nuzhdin SV. The fixation of malaria refractoriness in mosquitoes. Curr Biol 2004; 14(7):R264-265.
18. Macdonald. The epidemiology and control of malaria. London: Oxford University Press, 1957.
19. Brey PT. The impact of stress on insect immunity. Bull Institut Pasteur 1994:101-118.
20. Suwanchaichinda C, Paskewitz SM. Effects of larval nutrition, adult body size, and adult temperature on the ability of Anopheles gambiae (Diptera: Culicidae) to melanize sephadex beads. J Med Entomol 1998; 35:157-61.
21. Chun J, Riehle M, Paskewitz SM. Effect of mosquito age and reproductive status on melanization of sephadex beads in Plasmodium-refractory and -susceptible strains of Anopheles gambiae. J Invertebr Pathol 1995; 66:11-7.
22. Tabashnick WJ. Reflections on the anopheles gambiae genome sequence, transgenic mosquitoes and the prospect for controlling malaria and other vector borne diseases. J Med Entomol 2003; 40:597-606.
23. Lewontin R. The Triple Helix. Cambridge: Harvard University Press, 2000:136.
24. Boëte C, Paul REL, Koella JC. Reduced efficacy of the immune melanisation response in mosquitoes infected by malaria parasites. Parasitology 2002; 125:93-98.
25. Schwartz A, Koella JC. Melanization of Plasmodium falciparum and C-25 Sephadex beads by field caught Anopheles gambiae (Diptera: Culicidae) from Southern Tanzania. J Med Entomol 2002; 39:84-88.
26. Lyimo EO, Koella JC. Relationship between body size of adult Anopheles gambiae s.l. and infection with the malaria parasite Plasmodium falciparum. Parasitology 1992; 104:233-7.
27. Curtis CF, Pates HV, Takken W et al. Biological problems with the replacement of a vector population by Plasmodium-refractory mosquitoes. Parassitologia 1999; 41:479-81.
28. Zheng L, Wang S, Romans P et al. Quantitative trait loci in Anopheles gambiae controlling the encapsulation response against Plasmodium cynomolgi Ceylon. BMC Genetics 2003; 4.
29. Lambrechts L, Halbert J, Durand P et al. Host genotype by parasite genotype interactions underlying the resistance of anopheline mosquitoes to Plasmodium falciparum. Malaria J 2005; 4.
30. van Valen L. A New Evolutionary Law. Evolutionary Theory 1973; 1:1-30.
31. Gandon S, Mackinnon MJ, Nee S et al. Imperfect vaccines and the evolution of pathogen virulence. Nature 2001; 414:751-6.
32a. Boëte C, Paul RE. Can mosquitoes help to unravel the community structure of Plasmodium species? Trends Parasitol 2006; 22:21-25.
32b. Zambrano-Villa S, Rosales-Borjas D, Carrero J et al. How protozoan parasites evade the immune response. Trends Parasitol 2002; 18:272-278.
33. Bloom BR, Salgame P, Diamond B. Revisiting and revising suppressor T cells. Immunol Today 1992; 13:131-6.
34. Urban BC, Ferguson DJ, Pain A et al. Plasmodium falciparum-infected erythrocytes modulate the maturation of dendritic cells. Nature 1999; 400:73-7.
35. Urban BC, Roberts DJ. Malaria, monocytes, macrophages and myeloid dendritic cells: Sticking of infected erythrocytes switches off host cells. Curr Opin Immunol 2002; 14:458-465.
36. Modlin RL, Mehra V, Wong L et al. Suppressor T lymphocytes from lepromatous leprosy skin lesions. J Immunol 1986; 137:2831-4.

37. Strand MR, Pech LL. Immunological basis for compatibility in parasitoid-host relationships. Annu Rev Entomol 1995; 40:31-56.
38. Beckage NE. Modulation of immune responses to parasitoids by polydnaviruses. Parasitology 1998; 116:S57-64.
39. Vinson SB. How parasitoids deal with the immune system of their host: An overview. Archives of Insect Biochem Physiol 1990; 13:3-27.
40. Rizki RM, Rizki TM. Selective destruction of a host blood cell type by a parasitoid wasp. Proc Natl Acad Sci USA 1984; 81:6154-8.
41. Rizki RM, Rizki TM. Parasitoid virus-like particles destroy Drosophila cellular immunity. Proc Natl Acad Sci USA 1990; 87:8388-92.
42. Rizki TM, Rizki RM. Parasitoid-induced cellular immune deficiency in Drosophila. Ann N Y Acad Sci 1994; 712:178-94.
43. Sinden RE, Billingsley PF. Plasmodium invasion of mosquito cells: Hawk or dove? Trends Parasitol 2001; 17:209-11.
44. Boëte C, Paul REL, Koella JC. Direct and indirect immunosuppression by a malaria parasite in its mosquito vector. Proc R Soc Lond B Biol Sci 2004; 271:1611-1615.
45. Schmid-Hempel P. Variation in immune defence as a question of evolutionary ecology. Proc R Soc Lond B Biol Sci 2003; 270:357-66.
46. Moret Y, Schmid-Hempel P. Survival for immunity: The price of immune system activation for bumblebee workers. Science 2000; 290:1166-8.
47. Koella JC, Boëte C. A model for the coevolution of immunity and immune-evasion in vector-borne diseases, with implications for the epidemiology of malaria. Am Nat 2003; 161:698-707.
48. Boëte C. Malaria parasites in mosquitoes: laboratory models, evolutional temptation and the real world. Trends Parasitol 2005; 21:445-447.
49. Castro J. Millet D. Malaria and structural adjustment: Proof by contradiction. In: Boëte C, ed. Genetically Modified Mosquitoes for Malaria Control. Georgetown: Landes Bioscience, 2006; 2:16-23.

CHAPTER 8

The Epidemiological Consequences of Reducing the Transmission Intensity of *P. falciparum*

Hugh Reyburn* and Chris Drakeley

Abstract

Control methods which aim to reduce the human host's exposure to infected mosquitoes (as is the desired outcome of a release of transgenic mosquitoes) will ideally reduce the frequency of malarial infection. This change in transmission intensity will affect the development of immunity to malaria and the age at which infections are acquired. Both of these have important consequences for the overall burden of malarial disease with some suggestion that transmission reduction might not be beneficial in all scenarios. In this chapter we review the methods for measuring malaria transmission and summarise the data on disease epidemiology in these settings.

Introduction

Transgenic technology applied to malaria control has led to the idea of using genetically modified (GM) mosquitoes able to interrupt malaria transmission. While technological advances are being made in this area[1] there remains uncertainty as to how their introduction would affect the human burden of malaria. Such an impact will depend strongly on the efficiency of refractoriness[2] and on the preintervention levels of transmission.[3] In addition there are fundamental epidemiological questions relating to the level of population immunity and the consequent burden of malaria are common to any intervention designed to reduce exposure to infection and are the focus of this chapter.

At high levels of transmission there have been concerns that reducing transmission could result in a rebound in mortality in older age groups as the population level of immunity realigns to new levels of exposure. It has been argued that high rates of infection in early life may stimulate active immunity at a time when other mechanisms such as maternally acquired antibodies and the persistence of foetal haemoglobin reduce the risk of severe disease.[4] Other age-dependent effects may also operate, particularly that cerebral manifestations of malaria are relatively more common among older children than infants. These concerns were first raised in the context of the WHO campaign for the eradication of malaria in the 1950s[5,6] and have resurfaced with the growing use of insecticide treated nets (ITNs).[4]

At the other end of the transmission spectrum reductions from low and very low levels may result in a change from 'stable' to 'unstable' (i.e., epidemic-prone) malaria. There are

*Corresponding Author: Hugh Reyburn—Department of Infectious and Tropical Diseases, London School of Hygiene and Tropical Medicine, Keppel Street, London WC1E 7HT, U.K.; Joint Malaria Programme, KCMC, Moshi, Tanzania. Email: hugh.reyburn@lshtm.ac.uk

Genetically Modified Mosquitoes for Malaria Control, edited by Christophe Boëte.
©2006 Landes Bioscience.

Table 1. Classification of malaria transmission

Spleen Rate[1] (% palpable spleen in children age 2-9 yrs)	Parasite Prevalence[1] (% in children age 2-9 yrs)	EIR[2] (infectious bites per person per year)	Endemicity
1-10%	1-10%	<1	Hypoendemic
10-50%	10-50%	1-10	Mesoendemic
50-75%	50-75%	11-100	Hyperendemic
>75%	>75%	>100	Holoendemic

[1] Garnham PCC. Malaria parasites and other haemosporidia. Oxford: Blackwell, 1966; [2] Beier JC, Killeen GF, Githure JI. Short report: entomologic inoculation rates and Plasmodium falciparum malaria prevalence in Africa. Am J Trop Med Hyg 1999; 61(1):109-113.

few studies on which to base conclusions about this scenario but we consider the evidence later in the chapter.

The primary problem in resolving these questions, particularly at high transmission, is the difficulty in maintaining a controlled trial of an effective intervention for more than about two years, and this is insufficient time to observe rebound mortality. An alternative approach to prolonged controlled trials is to compare the burden of malaria in populations naturally exposed to different levels of transmission but this method has limitations.[7-10] Stable populations living under different levels of challenge have higher frequencies of genes that protect against malaria mortality and parasite diversity and possibly resistance to antimalarials may be greater at high transmission.[3] In addition, malaria itself may be the cause and result of economic differences which are likely to affect overall mortality due to its multiple effects on nutrition, access to and quality of care etc.[11] Finally, no ideal measure of exposure to *Plasmodium falciparum* (*P. falciparum*) currently exists that can reliably characterise the exposure of large populations that are needed if outcomes of severe malaria and mortality are to be compared.[12]

Measures of Malaria Transmission

Traditional methods of categorising transmission intensity of *P. falciparum* are summarised in Table 1. Each method has limitations that constrain current knowledge of how the human burden of malaria varies with transmission. However, in the context of GM mosquitoes, a combination of the approaches discussed below could provide improved measures of long-term exposure over large populations.

Entomological

The gold standard measure for (mosquito-man) transmission intensity is the entomological inoculation rate (EIR). The EIR is classically derived from the density of man-biting anopheline mosquitoes, sporozoite rate and the human blood index and represents the number of infectious bites an individual is likely to be exposed to over a defined period of time usually one year (ib/p/year). The human biting catch (HBC) is the most accurate method for assessing man biting rates though this technique is limited due to ethical and logistical constraints. Light trap catches (LTC) and pyrethrum spray catches (PSC) represent viable alternatives and have been evaluated against HBC.[13,14] Obtaining reliable and reproducible estimates of EIR is time consuming (and expensive) and subject to seasonal or meteorological fluctuations. Knowledge of local vectors and their feeding behaviour is clearly vital.

A review of EIR in Africa found there to be marked heterogeneity in malaria risk across the continent[15] from more than 2 infectious bites per night (884 ib/p/yr) to zero. Most reported studies have yet to consider confidence intervals (CI) around an EIR estimate which in areas of unstable and low malaria transmission are likely to be wide as the chances of catching an infected mosquito are extremely low. Any intervention which aims to reduce the proportion of infected mosquitoes (e.g., genetically modified mosquitoes) must be sufficiently powered to detect a significant reduction in this proportion pre and post intervention.[16]

Parasitological

Transmission estimates based on the prevalence of human infection (the parasite rate, PR) are informative as they represent actual (rather than potential) infections and are a direct measure of the disease within the community. In theory, sufficient numbers of people can quickly (and inexpensively) be tested (and retested) to obtain reliable estimates.

PR has been used to define malaria endemicity[17] and it has been used to classify malaria endemicity and recently has been correlated with EIR.[18] However there are similar caveats for PR and the EIR and relationships are not straightforward even in relatively small geographical locations;[12,19] in particularly PR is susceptible to micro heterogeneity caused by climatic factors and the socio-economic determinants of health seeking behaviour.

Clinical

A classical correlate of transmission intensity is the measurement of spleen rates in children.[20,21] Measurement of anaemia either as direct haemoglobin levels (Hb) or classification into the prevalence of mild or moderate forms has also been shown to correlate with transmission intensity.[12] The utility of these measures, Hb in particular, is that they are likely to be more reliable than a single blood film and will reflect longer term trends in exposure to malaria. However, different methods are often used to assess Hb levels and spleen rates are subject to observer variation. There are also potential confounders with other diseases which can cause both splenomegaly and/or anaemia, the latter may also be due to nutrient deficiency.

Serological

One possible alternative is to combine parasite prevalence with serological measurement of anti-parasite IgM.[22] This has been used to demonstrate reduced force of infection in bed net trials and subsequently used to characterise transmission intensity.[23] Other studies have shown contrasting results with no change in antibody levels in studies with treated curtains[24] and a reduction in antibody recognition with treated nets.[25] Similarly, contrasting data have come comparing EIR with responses to merozoite surface protein 1.[26,27] Clearly, both the choice of antigen (and the longevity of the antibody response it elicits) and study subjects are important. However the screening of large numbers of samples quickly and the potential for using more immunogenic antigens at lower transmission makes this approach attractive especially when blood taking can be combined with community health surveys.[27] One intriguing possibility is the measurement of anti-mosquito saliva antibodies in exposed populations (Mitchell, Drakeley, and Billingsley unpublished data). Total antibody levels may reflect overall exposure and specific isotypes (IgE and IgM) more recent exposure history. In theory, as the proteome of saliva becomes known, recombinant antigens could be used to refine identification to biting of specific species and reductions in vector host contact.

Geographical Information Systems

Recent advances in data collection techniques and remote sensing have led to increasingly advanced models for malaria transmission.[28-31] These are largely climate driven with additional information used to refine the model for a particular scenario: malaria risk,[28] vector distribution,[30] epidemic prediction[31] and urban malaria.[32] However GIS models may lack precision to define transmission at the local scale and a recent analysis suggested that they over-estimate those at no risk.[33] Modelling is becoming increasingly important in predicting malaria outbreaks in epidemic prone areas.[34]

The Effect of Reducing Transmission Intensity in Endemic Areas

Limitations in methodologies for measuring the transmission intensity of *P. falciparum* mean that transmission is routinely classified into broad bands (Table 1). Existing evidence has used these bands to predict the effects of reducing transmission on uncomplicated and severe malaria and how these are reflected in malaria-specific and all-cause mortality. As will be seen, these effects vary with transmission intensity and may change over time as the population level of immunity adjusts to new levels of exposure; concerns have been expressed that the result may not always be beneficial.[4]

Nonsevere Malaria

Malaria can present with a wide range of nonspecific symptoms, and generally it is not possible to discriminate between malaria and a number of other common illnesses. Non severe malaria is thus defined as a febrile illness with a positive test for malaria parasitaemia and no obvious alternative cause of illness.

Since, in endemic areas, asymptomatic parasitaemia is common many cases meeting the above definition are likely to be suffering from a nonmalarial illness, i.e., the definition of nonsevere malaria lacks specificity even when slide results are available. While the inclusion of the density of parasitaemia can provide some refinement to the definition, the problem of specificity remains.[35]

In controlled trials the use of ITNs has been associated with a reduction in the incidence of nonsevere malaria by almost 50%.[36] The sustainability of this reduction can be estimated by comparing stable populations living under different levels of *P. falciparum* challenge. In children under the age of 18 months naturally exposed to a 10-fold difference in EIR in Tanzania there was a 1.6 fold increase in the incidence of parasitaemia with fever in the high compared to the lower transmission area, suggesting that in infants and young children artificial reductions in transmission intensity are likely to result in a sustained reduction in nonsevere malaria.[37]

However, this effect may not be sustained among older children and adults. In two closely monitored villages in Senegal (Dielmo and Ndiop) exposed to EIR of 200 and 20 ib/p/yr respectively the age-specific incidence of any clinical episode of malaria was 'right shifted' in the low transmission village compared to the high transmission village. The lifetime number of malaria attacks, projected from population and malaria incidence data, was found to be 43 per person by the age of 60 years in the high transmission village and 62 per person by the age 60 in the low transmission village (Fig. 1).[38] Similarly in Kenya, Clarke et al found the incidence of such attacks in school children in the Kenyan highlands, an epidemic prone area, to exceed that in schoolchildren in a holoendemic area suggesting that the burden of nonsevere disease may remain comparable at all levels of malaria transmission.[39] The problems of accuracy of diagnosis and attributing the cause of febrile episodes make a definitive answer difficult as at high transmission levels the probability that febrile illness in the presence of parasitaemia but which are due to nonmalarial causes increases.[35]

Severe Malaria

The definition of severe malaria is based on the presence of *P. falciparum* parasitaemia, clinical features which have been empirically found to be associated with significant mortality, and the absence of an obvious alternative cause.[40] In African children the presence of one or more of 6 'syndromes' capture the large majority of severe and fatal cases of malaria.[41,42] These are:

1. Unrousable coma (a lower level of coma called 'impaired consciousness' is a supporting criterion)
2. Respiratory distress (abnormally deep breathing associated with in-drawing of the chest wall on inspiration).
3. Severe anaemia (Hb<5 g/dl)
4. Repeated convulsions

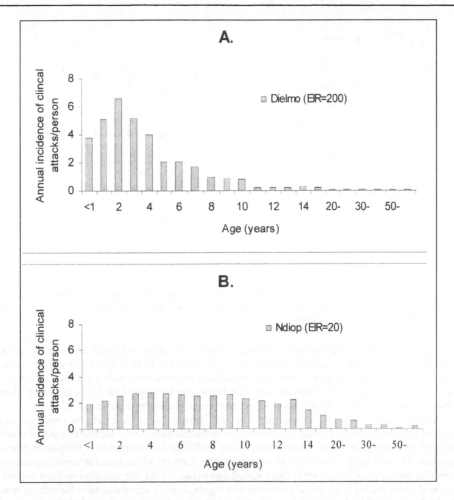

Figure 1. Annual incidence of malaria attacks by age in permanent residents of two Senegalese villages with markedly different *P. falciparum* transmission. A) Ndiop (EIR = 20 ib/p/y) and B) Dielmo (EIR = 200 ib/p/y). Reproduced with permission from: Trape JF, Rogier, C. Combating malaria morbidity and mortality by reducing transmission. Parasitol Today 1996; 12:236-240.

5. Hypoglycaemia
6. Prostration (inability to sit unsupported over the age of 8 months, or to suck if under the age of 8 months; prostration is technically a supporting criterion of severe malaria)

Case fatality is highest among children with coma, respiratory distress or hypoglycaemia and is strongly influenced by delay in treatment, the quality of care and presence of associated conditions such as dehydration or malnutrition. Case fatality of the specific syndromes varies from under 5% to over 25%, with children suffering from combinations of syndromes having a particularly high mortality. Overall 'severe malaria' is generally associated with a case fatality of between 5% and 10%.[41,42]

Community-based data on severe malaria lack sensitivity and specificity in distinguishing between malaria and other common infections[43,44] and hospital-based data remain the primary source of information on how both the incidence and severity of malaria is likely to change following reductions in transmission intensity.

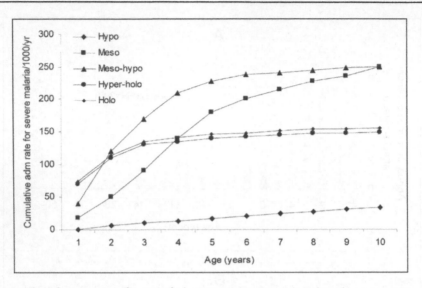

Figure 2. Cumulative age-specific rates of admission with severe malaria from five communities with increasing intensity of *P. falciparum* transmission. Reproduced with permission from Elsevier (The Lancet 1997; 349(9066):1650-1654).[23]

Trape et al (1987) were the first to compare hospital admission and death rates as a proxy for the longer term effects of malaria control. In a study based in Brazzaville he found little variation in both admission and death rates due to malaria between areas with a 100-fold difference in transmission intensity.[45] Although these findings may be atypical of Africa generally, they raised a serious public health question in the context of the wider deployment of ITNs.

A larger study to address this question compared hospital admission rates for severe malaria to 3 hospitals from five communities (two of the hospitals each served identifiable populations exposed to different levels of transmission) living at different intensities of *P. falciparum* transmission. As a marker for variations in hospital usage, data on admissions for acute respiratory infection were also collected; while these varied more than 2-fold, the variation was not systematic with transmission level. Results are shown in Figure 2; while admission rates for infants were the highest at the 2 high transmission sites, cumulative admission rates for malaria by the age of 10 years were highest among populations exposed to moderate to low transmission intensity.[23] Malaria admissions in the lowest transmission setting were consistently low at every age.

Inherent in the methodology of hospital-based studies is the potential for confounding (social, genetic, treatment seeking etc) when comparing admission rates between distant populations.[45] However, Schellenberg et al avoided many of these problems by studying the effect of a falling malaria transmission on admission rates for severe malaria in Ifakara town, Tanzania. In Ifakara between 1994 and 2001 age-specific parasite prevalence in children fell by at least 50%, the change being attributed at least in part to improved treatment (chloroquine was replaced by SP as first line treatment in 1999) and higher coverage with insecticide treated nets. During this time the incidence of nonsevere clinical malaria also fell by almost 50% in infants but the number of hospital admissions for malaria rose by 13% in spite of a fall in nonmalaria admissions (possibly related to the introduction of hospital charges).[46]

Three other studies have also suggested a similar 'saturation effect' of admission rates for severe malaria which occurs somewhere between exposure levels of 1 and 30 ib/p/yr.[19,47,48] Below this point of saturation Snow's data suggest that the incidence of severe malaria in areas of moderate and low endemicity (PR<25% in children under 10 years of age) tends to increase linearly with transmission intensity suggesting the greatest relative reductions in the incidence of severe malaria can be achieved by reducing transmission intensity in areas of moderate and low transmission.[49]

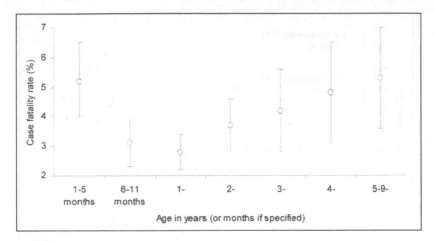

Figure 3. Age-specific case fatality rates of 9,300 paediatric admissions for malaria to 10 African hospitals. Reproduced with permission from: Marsh K, Snow RW. Malaria transmission and morbidity. Parassitologia 1999; 41(1-3):241-246, ©1999 Lomardo Editore.

Variation in Case Fatality of Severe Malaria with Transmission Intensity

As outline above cerebral malaria is associated with a case fatality rate that is 2-3 times that of severe anaemia.[41,42] In areas of high transmission, severe anaemia among infants and young children dominates the clinical picture of severe malaria, and this is also reflected in community surveys where anaemia, particularly at high transmission, is strongly age-dependent.[50] In areas of low, and especially seasonal, transmission cerebral malaria becomes relatively more common although comparisons of the incidence of cerebral malaria between sites are complicated by the lack of a uniform definition and that the cerebral manifestations of malaria are due to a variety of pathological processes.[51] A consistent and related finding is that within any given area the age of children with cerebral malaria is older than that of children with severe anaemia.[23,48,52-54]

The above would suggest that case fatality rates due to severe malaria is likely to increase with age and/or falling transmission intensity but there are few data to confirm this. Data from surveillance of 9,300 admissions to 10 hospitals in sub-Saharan Africa collected between 1990 and 1997 (Fig. 3) suggest high fatality under the age of 6 months (an age under which severe malaria is relatively uncommon), relatively low case fatality around the age of 1 year and a subsequent progressive rise.[55]

All-Cause Child Mortality and Transmission Intensity

If the incidence and case fatality of severe malaria reach a plateau, or even decline, with increasing transmission then it follows that partial control of malaria in areas of high transmission will not result in long-term gains. However, this conclusion may well not be valid for primarily two reasons.

Firstly, hospital admission rates on which the above conclusions are largely built may be biased towards illness in older children who are more able than infants to draw attention to their symptoms, especially if these relate to the insidious onset of severe malarial anaemia. Secondly, malaria may be an important indirect cause of mortality, i.e., intense transmission leads to a state of chronic or frequent parasitisation that increases the risk of death in young children from a variety of causes.[8,53,56] The evidence for this is that in controlled trials of ITNs[57] and chemoprophylaxis[58] all-cause mortality was reduced by about twice as much as would be expected from baseline estimates of malaria-specific mortality. Historical declines in all cause mortality associated with 'pseudo-eradication' of malaria in Sri Lanka were also

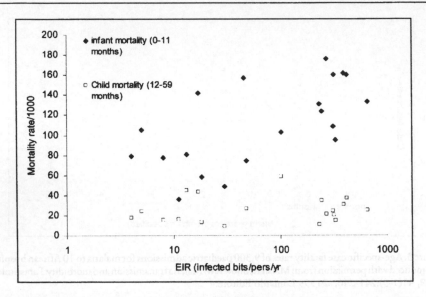

Figure 4. African all-cause child and infant mortality rates by entomological inoculation rate for *P. falciparum*. Reproduced with permission from: Smith TA, Leuenberger R, Lengeler C. Child mortality and malaria transmission intensity in Africa. Trends Parasitol 2001; 17(3):145-149.

greater than expected from known malaria-specific mortality, even after allowing for the existing background trend in falling mortality associate with development.[59-61] The indirect contribution of malaria to all-cause mortality is likely to be mitigated by the fact that in high transmission areas, children surviving infancy become more rapidly immune to the direct effects of malaria.

Malaria as a risk factor rather than a direct cause of mortality is more likely to exert an effect among young children living at high transmission for several reasons; low birth weight and neonatal anaemia are both more common in areas of high transmission,[62] the risks of anaemia are maximal in the first year of life for physiological and nutritional reasons, and bacteraemia is most common among infants under the age of 6 months. These effects will be enhanced by the fact that signs of illness in young babies are harder to detect than in older children and thus infants may present later for treatment than older children.

The importance of malaria as an indirect cause of mortality is likely to be reflected in all-cause childhood mortality. Smith et al have thus examined rates of all-cause mortality among infants and children age between 12 and 60 months of age recorded from 20 DSS studies since 1980 (Fig. 4). There was a significant variation between infant mortality and EIR, but mortality in older children varied less and there was no clear trend with EIR.[45,63]

A similar analysis used of the parasite ratios in children (<15 years of age) categorised into bands (<25%, 25-50%, 51-74% and >74%) rather than EIR.[3] Initial results suggested that at all transmission settings where the parasite rate was greater than 25% similar levels of mortality among children under the age of 5 years were experienced. However, a recent extension of this analysis showed a positive relationship between prevalence and mortality extending to prevalence's of 50% though plateauing at the highest prevalences.[49]

Direct Measures of Effectiveness of Malaria Control on All Cause Mortality

Of 18 controlled trials that have been conducted on ITNs only two have been in areas of high transmission and in these the protective efficacy appeared lower than in areas of moderate or low transmission.[49,64,65] Lengeler et al have argued for a focus on attributable risk (the number of lives saved) rather than rate ratios (protective efficacy), and when reanalysed in this

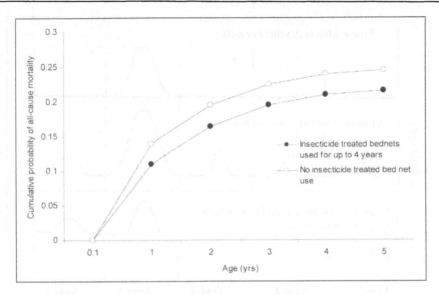

Figure 5. Cumulative probability of all-cause mortality by age for children with up to 4 years of insecticide-treated bednet use compared with children with no insecticide treated bednet use from synthetic cohorts of summed age-specific death rates. Reproduced with permission from: Lindblade KA, Eisele TP, Gimnig JE et al. Sustainability of reductions in malaria transmission and infant mortality in western Kenya with use of insecticide-treated bednets: 4 to 6 years of follow-up. JAMA 2004; 291(21):2571-2580.

way there was little difference in the number of lives saved among children under the age of 5 years in a wide range of transmission intensities.[66]

Of more concern is the possibility that in high transmission settings protective efficacy in older children might decline or even reverse over time.[65] However, this has so far not proved to be the case; in the two studies mentioned above, the gains in mortality that had been made in the first 1-2 year of the trials were not associated with a rebound in mortality over 6 and 7 years of follow-up respectively, although these were not in direct comparison with controls.[64,67] Recently a more detailed follow-up for 4 years following the end of a 2-year controlled trial of ITNs in an area of intense perennial transmission in Western Kenya has provided results that are both reassuring and consistent with analyses of all cause child mortality at varying transmission. ITNs had a protective efficacy of 22% against all cause mortality in children aged 1-11 months in the intervention villages during 2 years of the trial (but with no effect on mortality in older children), and throughout 4 years of follow-up there was no evidence of increased mortality in children aged 1-5 years old (Fig 5).[68]

The Effect of Transmission Reduction in Areas of Unstable Malaria

The sections above have considered the epidemiology of malaria in areas where transmission is sufficiently high to induce protective immunity during childhood and the population burden of disease varies little from year to year.

Malaria epidemics occur where exposure is insufficient to produce a significant population level of immunity to severe disease and thus an abrupt increase in transmission can result in a malaria epidemic. However, the pattern of such 'epidemics' depends strongly on a variety of conditions schematically summarised in Figure 6.

The causes of epidemics can be broadly divided into those that result from unusually wet and warm seasonal rains (e.g., associated with the El-Niño southern oscillation), a gradual breakdown of successful malaria control strategies or population movement and environmental change often linked to complex emergencies.

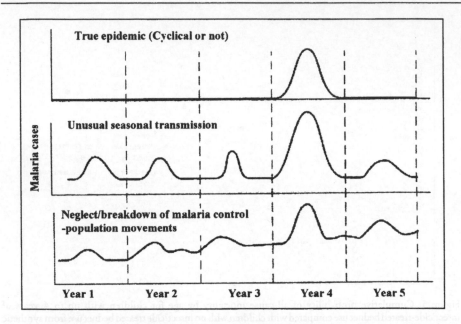

Figure 6. Classification of major epidemic types. The top panel shows the pattern of cases in a true epidemic, the middle panel shows how cases may occur in epidemics caused by unusual seasonal transmission and the bottom panel shows epidemics caused by a neglect or breakdown in malaria control and/or population movements. Reproduced with permission from: Roll back malaria: Report of technical working group on the prevention and control of malaria epidemics. WHO/CDS/RBM/2002.4.

Epidemics are inherently unpredictable in frequency duration and severity and for these reasons it is difficult to make confident statements about their likelihood following malaria control. Up to 25% of the population in Sub-Saharan Africa lives in epidemic prone areas[69] and very few of these are the result of deliberate malaria control activities; conversely, most areas where malaria was eradicated have not experienced epidemics. However, a minority of epidemics are the result of malaria control activities whose effectiveness has broken down. Examples are the severe epidemics in Sri Lanka, India and the highlands of Madagascar following the breakdown of residual spraying in the 1960s.[70]

Whatever the underlying cause, malaria epidemics have, to at least some extent, certain characteristics in common. Firstly, all ages tend to be affected, although children often have the highest case fatality suggesting that immunity to the most severe forms of malaria may be acquired after relatively few infections and may be retained for many years.[71] Secondly, health services are often overwhelmed and case fatality is often significantly higher than in areas of stable transmission.[70] Thirdly, there is often significant disruption of local infrastructure with a consequent reduction in the capacity to make a prompt and effective response.[72,73]

The increased vulnerability of pregnant women to malaria varies in its manifestations between stable endemic and unstable transmission areas. In areas with stable endemic malaria the risk of severe malaria and low birth weight are generally confined to the first pregnancy while in low transmission areas this risk continues into subsequent pregnancies.[74] Premature delivery is more common in epidemic prone areas while intra-uterine growth retardation is more common in areas of stable transmission.[75]

The epidemiological consequences of reducing *P. falciparum* transmission through the use of transgenic mosquitoes in areas of low transmission are still unclear. If there is an influx of infectious human hosts then refractory mosquitoes would abrogate the spread. However, if there is a recovery or reintroduction of a efficient vector then epidemics seem likely, as for example seen in the resurgence in *An. funestus* which was implicated as the reason for malaria epidemics in the Madagascan highlands.[72]

In low transmission settings new tools are emerging to monitor trends in malaria and to provide early warning of epidemics. In the presence of a large number of uncertainties regarding the long term effectiveness of transgenic mosquitoes it is particularly important to use these methods of monitoring and epidemic early warning for extended periods in areas where they are deployed.[73,76,77]

Conclusions

What can be concluded from the existing data on the likely effect of reducing transmission intensity of *P. falciparum* in Africa?

- Firstly that reducing malaria transmission will result in reductions in the incidence of both nonsevere, severe and fatal malaria in the short term. (i.e., 1-2 years following malaria control) and that these reductions are sustained among infants.

- Secondly, reductions in the transmission intensity of *P. falciparum* are likely to achieve the greatest *relative* reduction in severe and fatal malaria in areas of low and moderate transmission intensity. However, the number of lives saved through such reductions (attributable mortality) is likely to be similar over a wide range of transmission settings.

- Thirdly that the evidence for rebound mortality among older children following reductions in transmission is derived from hospital-based studies which is not entirely consistent with comparisons of all cause mortality following reductions in transmission. The explanation is likely to be due to the tendency of hospital based data to under-estimate the burden of malaria in infants compared to older children, and by a significant degree of indirect mortality due to malaria in high transmission settings.

- Fourthly that reducing transmission intensity in low transmission areas will require additional efforts for long term monitoring for epidemics.

It is obviously essential that public health measures such as reductions in the transmission intensity of *P. falciparum* are evaluated for their costs and benefits, and particularly for any potential for long-term harm. Such a process continues with respect to malaria, but over the last few years a broad consensus has been reached of the benefits of deploying the available means, and to actively seek new means, to reduce transmission of *P. falciparum*.

The novelty of GM mosquitoes as a tool to reduce malaria transmission highlights the particular need to ensure that close and sustained monitoring following their introduction.

References

1. Catteruccia F, Godfray HC, Crisanti A. Impact of genetic manipulation on the fitness of Anopheles stephensi mosquitoes. Science 2003; 299(5610):1225-1227.
2. Boëte C, Koella JC. A theoretical approach to predicting the success of genetic manipulation of malaria mosquitoes in malaria control. Malar J 2002; 1(1):3.
3. Snow RW, Marsh K. The consequences of reducing transmission of Plasmodium falciparum in Africa. Adv Parasitol 2002; 52:235-264.
4. Snow RW, Marsh K. Will reducing Plasmodium falciparum transmission alter malaria mortality among African children. Parasitol Today 1995; 11(5):188-190.
5. Wilson DB. On the present and future malaria outlook in East Africa. East Afr Med J 1949; 26:378-385.
6. Wilson DB, Garnham PCC, Swellengrebel NH. A review of hyperendemic malaria. Trop Dis Bull 1950; 47:677-698.
7. D'Alessandro U, Coosemans M. Concerns on long-term efficacy of an insecticide-treated bednet programme on child mortality. Parasitol Today 1997; 13(3):124-125.
8. Greenwood BM. Malaria transmission and vector control. Parasitol Today 1997; 13(3):90-92.
9. Lengeler C, Smith TA, Armstrong Schellenberg J. Focus on the effect of bednets on malaria morbidity and mortality. Parasitol Today 1997; 13(3):123-124.
10. Molineaux L. Nature's experiment: What implications for malaria prevention? Lancet 1997; 349(9066):1636-1637.
11. Sachs J, Malaney P. The economic and social burden of malaria. Nature 2002; 415(6872):680-685.

12. Drakeley CJ, Carneiro I, Reyburn H et al. Altitude-dependent and -independent variations in Plasmodium falciparum prevalence in northeastern Tanzania. J Infect Dis 2005; 191:1589-1598.
13. Lines J, Curtis C, Wilkes T et al. Monitoring human biting mosquitoes in Tanzania with light traps hung beside mosquito nets. Bull Entomol Res 1991; 81:77-84.
14. Mbogo CN, Glass GE, Forster D et al. Evaluation of light traps for sampling anopheline mosquitoes in Kilifi, Kenya. J Am Mosq Control Assoc 1993; 9(3):260-263.
15. Hay SI, Rogers DJ, Toomer JF et al. Annual Plasmodium falciparum entomological inoculation rates (EIR) across Africa: Literature survey, Internet access and review. Trans R Soc Trop Med Hyg 2000; 94(2):113-127.
16. Drakeley C, Schellenberg D, Kihonda J et al. An estimation of the entomological inoculation rate for Ifakara: A semi-urban area in a region of intense malaria transmission in Tanzania. Trop Med Int Health 2003; 8(9):767-774.
17. Metselaar D, Van Thiel P. Classification of Malaria. Trop Geogr Med 1959; 11:157-161.
18. Beier JC, Killeen GF, Githure JI. Short report: Entomologic inoculation rates and Plasmodium falciparum malaria prevalence in Africa. Am J Trop Med Hyg 1999; 61(1):109-113.
19. Mbogo CN, Snow RW, Khamala CP et al. Relationships between Plasmodium falciparum transmission by vector populations and the incidence of severe disease at nine sites on the Kenyan coast. Am J Trop Med Hyg 1995; 52(3):201-206.
20. Gillies HM. Epidemiology of Malaria. In: Gillies HM, Warrell DA, eds. Bruce-Chwatts Essential Malariology. Boston: Edward Arnold, 1993:132-136.
21. Metselaar D. Spleens and holoendemic malaria in West New Guinea. Bull World Health Organ 1956; 15(3-5):635-649.
22. Snow RW, Molyneux CS, Warn PA et al. Infant parasite rates and immunoglobulin M seroprevalence as a measure of exposure to Plasmodium falciparum during a randomized controlled trial of insecticide-treated bed nets on the Kenyan coast. Am J Trop Med Hyg 1996; 55(2):144-149.
23. Snow RW, Omumbo JA, Lowe B et al. Relation between severe malaria morbidity in children and level of Plasmodium falciparum transmission in Africa. Lancet 1997; 349(9066):1650-1654.
24. Bolad A, Nebie I, Esposito F et al. The use of impregnated curtains does not affect antibody responses against Plasmodium falciparum and complexity of infecting parasite populations in children from Burkina Faso. Acta Trop 2004; 90(3):237-247.
25. Askjaer N, Maxwell C, Chambo W et al. Insecticide-treated bed nets reduce plasma antibody levels and limit the repertoire of antibodies to Plasmodium falciparum variant surface antigens. Clin Diagn Lab Immunol 2001; 8(6):1289-1291.
26. Singer LM, Mirel LB, ter Kuile FO et al. The effects of varying exposure to malaria transmission on development of antimalarial antibody responses in preschool children. XVI. Asembo Bay Cohort Project. J Infect Dis 2003; 187(11):1756-1764.
27. Drakeley CJ, Corran PH, Coleman PG et al. Estimating medium- and long-term trends in malaria transmission using serological markers of malaria exposure. Proc Natl Acad Sci USA 2005; 102:5108-5113.
28. Craig MH, Snow RW, le Sueur D. A climate-based distribution model of malaria transmission in sub-Saharan Africa. Parasitol Today 1999; 15(3):105-111.
29. Hay SI, Tucker CJ, Rogers DJ et al. Remotely sensed surrogates of meteorological data for the study of the distribution and abundance of arthropod vectors of disease. Ann Trop Med Parasitol 1996; 90:1-19.
30. Kiszewski A, Mellinger A, Spielman A et al. A global index representing the stability of malaria transmission. Am J Trop Med Hyg 2004; 70(5):486-498.
31. Zhou G, Minakawa N, Githeko AK et al. Association between climate variability and malaria epidemics in the East African highlands. Proc Natl Acad Sci USA 2004; 101(8):2375-2380.
32. Caldas de Castro M, Yamagata Y, Mtasiwa D et al. Integrated urban malaria control: A case study in Dar es Salaam, Tanzania. Am J Trop Med Hyg 2004; 71(2 Suppl):103-117.
33. Omumbo JA, Hay SI, Guerra CA et al. The relationship between the Plasmodium falciparum parasite ratio in childhood and climate estimates of malaria transmission in Kenya. Malar J 2004; 3(1):17.
34. de Savigny D, Binka F. Monitoring future impact on malaria burden in sub-saharan Africa. Am J Trop Med Hyg 2004; 71(2 Suppl):224-231.
35. Smith T, Schellenberg JA, Hayes R. Attributable fraction estimates and case definitions for malaria in endemic areas. Stat Med 1994; 13(22):2345-2358.
36. Lengeler C. Insecticide-treated bednets and curtains for preventing malaria. Cochrane Database Syst Rev 2000; (2):CD000363.

37. Smith T, Charlwood JD, Kitua AY et al. Relationships of malaria morbidity with exposure to Plasmodium falciparum in young children in a highly endemic area. Am J Trop Med Hyg 1998; 59(2):252-257.
38. Trape JF, Rogier C. Combating malaria morbidity and mortality by reducing transmission. Parasitol Today 1996; 12:236-240.
39. Clarke SE, Brooker S, Njagi JK et al. Malaria morbidity among school children living in two areas of contrasting transmission in western Kenya. Am J Trop Med Hyg 2004; 71(6):732-738.
40. WHO. Severe falciparum malaria. Trans R Soc Trop Med Hyg 2000; 94(supplement 1):1-2.
41. Schellenberg D, Menendez C, Kahigwa E et al. African children with malaria in an area of intense Plasmodium falciparum transmission: Features on admission to the hospital and risk factors for death. Am J Trop Med Hyg 1999; 61(3):431-438.
42. Marsh K, Forster D, Waruiru C et al. Indicators of life-threatening malaria in African children. N Engl J Med 1995; 332(21):1399-1404.
43. Snow RW, Basto de Azevedo I, Forster D et al. Maternal recall of symptoms associated with childhood deaths in rural east Africa. Int J Epidemiol 1993; 22(4):677-683.
44. Snow RW, Armstrong JR, Forster D et al. Childhood deaths in Africa: Uses and limitations of verbal autopsies. Lancet 1992; 340(8815):351-355.
45. Smith TA, Leuenberger R, Lengeler C. Child mortality and malaria transmission intensity in Africa. Trends Parasitol 2001; 17(3):145-149.
46. Schellenberg D, Menendez C, Aponte J et al. The changing epidemiology of malaria in Ifakara Town, southern Tanzania. Trop Med Int Health 2004; 9(1):68-76.
47. Trape JF, Zoulani A, Quinet MC. Assessment of the incidence and prevalence of clinical malaria in semi-immune children exposed to intense and perennial transmission. Am J Epidemiol 1987; 126(2):193-201.
48. Snow RW, Bastos de Azevedo I, Lowe BS et al. Severe childhood malaria in two areas of markedly different falciparum transmission in east Africa. Acta Trop 1994; 57(4):289-300.
49. Snow RW, Korenromp EL, Gouws E. Pediatric mortality in Africa: Plasmodium falciparum malaria as a cause or risk? Am J Trop Med Hyg 2004; 71(2 Suppl):16-24.
50. Schellenberg D, Schellenberg JR, Mushi A et al. The silent burden of anaemia in Tanzanian children: A community-based study. Bull World Health Organ 2003; 81(8):581-590.
51. Taylor TE, Fu WJ, Carr RA et al. Differentiating the pathologies of cerebral malaria by postmortem parasite counts. Nat Med 2004; 10(2):143-145.
52. Brewster DR, Greenwood BM. Seasonal variation of paediatric diseases in The Gambia, west Africa. Ann Trop Paediatr 1993; 13(2):133-146.
53. Greenwood BM. The epidemiology of malaria. Ann Trop Med Parasitol 1997; 91(7):763-769.
54. Reyburn H, Mbatia R, Drakeley C et al. Association of transmission intensity and age with clinical manifestations and case fatality of severe Plasmodium falciparum malaria. JAMA 2005; 293(12):1461-1470.
55. Marsh K, Snow RW. Malaria transmission and morbidity. Parassitologia 1999; 41(1-3):241-246.
56. Molineaux L. Malaria and mortality: Some epidemiological considerations. Ann Trop Med Parasitol 1997; 91(7):811-825.
57. Alonso PL, Lindsay SW, Armstrong Schellenberg JR et al. A malaria control trial using insecticide-treated bed nets and targeted chemoprophylaxis in a rural area of The Gambia, west Africa. 6. The impact of the interventions on mortality and morbidity from malaria. Trans R Soc Trop Med Hyg 1993; 87(Suppl 2):37-44.
58. Greenwood BM, Greenwood AM, Bradley AK et al. Comparison of two strategies for control of malaria within a primary health care programme in the Gambia. Lancet 1988; 1(8595):1121-1127.
59. Bradley D. Morbidity and Mortality at PareTaveta, Kenya and Tanzania, 1954-1966: The Effects of a Period of Malaria Control. Disease and Mortality in sub Saharan Africa. Oxford University Press, 1991:248-261.
60. Bradley D. Malaria: Old infections, changing epidemiology. Health Transition Review 2 1992; Supplement:137-153.
61. Molineaux L. The impact of parasitic diseases and their control with an emphasis on malaria in Africa. Health Policy, social policy and mortality prospects. Liege: Ordina Editions, 1985:13-44.
62. Goodman CA, Coleman PG, Mills AJ. Cost-effectiveness of malaria control in sub-Saharan Africa. Lancet 1999; 354(9176):378-385.
63. Smith T, Killeen G, Lengeler C et al. Relationships between the outcome of Plasmodium falciparum infection and the intensity of transmission in Africa. Am J Trop Med Hyg 2004; 71(2 Suppl):80-86.
64. Binka FN, Hodgson A, Adjuik M et al. Mortality in a seven-and-a-half-year follow-up of a trial of insecticide-treated mosquito nets in Ghana. Trans R Soc Trop Med Hyg 2002; 96(6):597-599.

65. Habluetzel A, Diallo DA, Esposito F et al. Do insecticide-treated curtains reduce all-cause child mortality in Burkina Faso? Trop Med Int Health 1997; 2(9):855-862.
66. Lengeler C, Armstrong-Schellenberg J, D'Alessandro U et al. Relative versus absolute risk of dying reduction after using insecticide-treated nets for malaria control in Africa. Trop Med Int Health 1998; 3(4):286-290.
67. Diallo DA, Cousens SN, Cuzin-Ouattara N et al. Child mortality in a West African population protected with insecticide-treated curtains for a period of up to 6 years. Bull World Health Organ 2004; 82(2):85-91.
68. Lindblade KA, Eisele TP, Gimnig JE et al. Sustainability of reductions in malaria transmission and infant mortality in western Kenya with use of insecticide-treated bednets: 4 to 6 years of follow-up. JAMA 2004; 291(21):2571-2580.
69. Worrall E, Rietveld A, Delacollette C. The burden of malaria epidemics and cost-effectiveness of interventions in epidemic situations in Africa. Am J Trop Med Hyg 2004; 71(2 Suppl):136-140.
70. Kiszewski AE, Teklehaimanot A. A review of the clinical and epidemiologic burdens of epidemic malaria. Am J Trop Med Hyg 2004; 71(2 Suppl):128-135.
71. Deloron P, Chougnet C. Is Immunity to malaria really short-lived? Parasitol Today 1992; 8(11):375-378.
72. Mouchet J. Origin of malaria epidemics on the plateaux of Madagascar and the mountains of east and south Africa. Bull Soc Pathol Exot 1998; 91(1):64-66.
73. Matola YG, Mwita U, Masoud AE. Malaria in the islands of Zanzibar and Pemba 11 years after the suspension of a malaria eradication programme. Cent Afr J Med 1984; 30(5):91-92, 95-96.
74. Nosten F, ter Kuile F, Maelankirri L et al. Malaria during pregnancy in an area of unstable endemicity. Trans R Soc Trop Med Hyg 1991; 85(4):424-429.
75. Sullivan AD, Nyirenda T, Cullinan T et al. Malaria infection during pregnancy: Intrauterine growth retardation and preterm delivery in Malawi. J Infect Dis 1999; 179(6):1580-1583.
76. Matola YG, Magayuka SA. Malaria in the Pare area of Tanzania. V. Malaria 20 years after the end of residual insecticide spraying. Trans R Soc Trop Med Hyg 1981; 75(6):811-813.
77. Sama W, Killeen G, Smith T. Estimating the duration of Plasmodium falciparum infection from trials of indoor residual spraying. Am J Trop Med Hyg 2004; 70(6):625-634.

CHAPTER 9

Malaria Parasite Virulence in Mosquitoes and Its Implications for the Introduction and Efficacy of GMM Malaria Control Programmes

Heather Ferguson,* Sylvain Gandon, Margaret Mackinnon and Andrew Read

Abstract

Initial scepticism about the ecological feasibility of the genetically modified mosquito (GM) approach for malaria control[1,2] has been supported by some recent experimental studies indicating that the insertion of transgenes, including those that induce refractoriness to malaria, confers a fitness cost to mosquitoes.[3-5] However, consideration of the possible fitness advantages of not becoming infected is also required to evaluate the net fitness of transgenic mosquitoes when introduced into natural populations. Therefore knowledge of whether malaria parasites are virulent to their vectors, and if so, to what magnitude, has direct relevance for forecasting the success of the GM approach. Here we summarize all known detrimental effects of malaria parasites on their mosquito vectors, and discuss their implications to the introduction of malaria-refractory genes in nature. Furthermore we review the mode of action by which transgenes generate refractoriness, and speculate on the evolutionary responses of *Plasmodium* to this killing mechanism. Finally, the virulence implications of current candidate GM phenotypes, both to mosquitoes and humans, are discussed.

Introduction

Initial scepticism about the ecological feasibility of the genetically modified mosquito (GM) approach for malaria control[1,2] has been supported by some recent experimental studies indicating that the insertion of transgenes, including those that promote resistance to malaria (*Plasmodium* sp.), confers a fitness cost to mosquitoes.[3-5] This suggests that the GM approach would have limited epidemiological impact, as the genes carrying refractoriness may not reach sufficiently high prevalence in vector populations to reduce disease transmission (see Box 1 for a description of the epidemiological impact of GM). Some argue that the low fitness of GM mosquitoes need not impede the spread of the refractory genes they carry as long as they are linked to an efficient genetic drive mechanism.[6] Certainly genetic drive will increase the rate of gene invasion, but only if a sufficiently high number of inseminations occur in the first place;

*Corresponding Author: Heather Ferguson—Public Health Entomology Unit, Ifakara Health Research and Development Centre, Tanzania; Laboratory of Entomology, Wageningen University, Wageningen, The Netherlands. Email: heather.ferguson@wur.nl

Genetically Modified Mosquitoes for Malaria Control, edited by Christophe Boëte. ©2006 Landes Bioscience.

which may not happen if the fitness costs of refractoriness are very high. Furthermore, given that no appropriate drive mechanism has yet been identified, and that the ability to tightly link an immune effector gene to a drive mechanism is in question,[7,8] the importance of mosquito fitness to GM control remains paramount.

However, by evading *Plasmodium* infection, GM mosquitoes would avoid one potential fitness cost to which wild-type mosquitoes are vulnerable. If *Plasmodium* prevalence and the cost of becoming infected is sufficiently high, it is possible that fitness benefits from never becoming infected could partially compensate for those associated with engineered refractoriness, and increase the likelihood of these genes invading wild mosquito populations. The term refractoriness implies resistance to infection, which in the case of GM mosquitoes can include both the total inability to develop any parasites (infection blocking), or a reduced probability of acquiring infection and harbouring high parasite burdens (infection reducing). In practice, refractoriness means only a reduced probability of infection and lower parasite burdens, as none of the transgenically-refractory mosquito strains currently available are completely resistant to *Plasmodium*.[9,10] If malaria parasites are virulent to their vectors, as defined by a reduction in fecundity and/or survival accompanying infection, it is possible that the fitness advantages of avoiding infection could help refractory genes invade into mosquito populations. Theoretical models illustrate that at least in principle, the rate at which refractory genes spread through a mosquito population should increase with parasite virulence.[6] This is because as a parasite becomes increasingly harmful to its host, the fitness benefits of enhanced resistance are greater. Knowledge of whether malaria parasites are virulent to their vectors, and if so, to what magnitude, thus has direct relevance for forecasting the success of the GM approach.

In addition to its epidemiological relevance, knowledge of parasite virulence in mosquitoes is critical for prediction of the direction under which malaria parasites could evolve under GM pressure. Parasites and their hosts are locked in a co-evolutionary battle where resistance on the part of the host is continually broken down by new invasion strategies on the part of the parasite.[11] To date, most studies investigating the efficacy of malaria refractory genes have considered their ability to block only one parasite genotype from developing in one mosquito genotype, with the average efficiency of resistance genes within genetically diverse populations being unknown. Studies of drug resistance in human populations, and insecticide resistance in mosquitoes, indicate that both malaria parasites and their vectors harbour extensive genetic diversity, and can mount rapid evolutionary responses to control measures. Thus it is likely that *Plasmodium* could evolve to circumvent the defences of a refractory mosquito within a short period of time. The consequences of resistance evolution could be considerably more severe than simple erosion of GM control efficacy if this process generates selection for parasites that are not only more virulent to mosquitoes, but also to humans. Here we summarize all known detrimental effects of malaria parasites on mosquitoes, and discuss their implications to the introduction of malaria-refractory genes in nature. We focus both on the epidemiological importance of *Plasmodium* virulence to mosquitoes to a GM strategy (e.g., is there a fitness cost to infection that would be avoided in refractory mosquitoes?), and discuss possible downstream evolutionary consequences for both vectors and humans should the strategy be successful.

Are Malaria Parasites Virulent to Their Vectors?

There is a long-standing assumption that parasites should evolve towards avirulence in their vector in order to increase their chance of being transmitted before vector death. However, most vector-borne parasites replicate within their vector, a process that may necessarily inflict some degree of virulence.[12] *Plasmodium* requires at least 10 days of growth within the *Anopheles* mosquitoes that transmit it before it can re-infect a new vertebrate host. During this period, parasites sexually reproduce, undergo four distinct life-history transitions (gamete, ookinete, oocyst and sporozoite) and can multiply more than 1000-fold.[13] It is possible that this process consumes mosquito resources or causes damage that results in a net decrease in lifetime fitness. This virulence could be manifested either as a reduction in mosquito fecundity or longevity. From an epidemiological perspective, the latter is deemed to have a more dramatic effect on

Table 1. Mean effect sizes (r) of Plasmodium on mosquito survival (calculated using the program METAWIN)[77] derived from meta-analysis of 11 laboratory studies (summarizing 24 experiments) described by Ferguson and Read[16]

Sample	Mean r	95% CI of r	N
All experiments	0.287	0.136—0.470	24
Studies (experiments pooled within a study)	0.259	0.102—0.447	11
Experiments of natural associations	0.061	-0.004—0.170	9
Experiments of nonnatural associations	0.436	0.201—0.705	13
Experiments ending before sporozoite invasion	0.129	0.055—0.218	10
Experiments ending after sporozoite invasion	0.395	0.147—0.664	14

If there was no effect of malaria infection on longevity, effect size would be zero; positive effect sizes indicate mortality was increased by infection. Confidence intervals (CI) were obtained by bootstrapping. Statistical analysis was conducted on Z-transformed values of r. Reprinted from: Ferguson HM, Read A. Why is the effect of malaria parasites on mosquito survival still unresolved? Reprinted with permission from Trends in Parasitology 18:259. ©2002 Elsevier.

parasite transmission than the former because small decreases in mosquito survival strongly reduce parasite transmission opportunities,[14] whereas a drop in mosquito fecundity does not.[15] Evidence indicates that *Plasmodium* can reduce both the fecundity and survival of its vector.

Survival

Ferguson and Read[16] conducted a meta-analysis of laboratory studies in which mosquitoes were experimentally infected with malaria, and their subsequent survival monitored. Of 22 studies reviewed, the proportion reporting a detrimental effect of *Plasmodium* was similar to that for which no effect was found (41% vs. 59%). For all accounts of reduced survival to have arisen by chance alone (Type 1 errors), there would need to be about 360 unpublished studies with null results, as well as nine showing increased survival. Given the experimental effort involved in survival studies, and the novelty of showing that malaria is a longevity-enhancer, this degree of under reporting seems unlikely. Survival reduction, thus, appears to be a genuine although not universal outcome. This analysis indicated that observations of virulence were linked to experimental design, with reduced survival being much more likely in studies using unnatural vector-parasite associations, and in those that monitored survival until the sporozoite stage of parasite development and beyond (Table 1). These results appear to support the notion that *Plasmodium* is harmful only in novel vector species, an idea frequently proposed to explain the lack of virulence in studies of natural infections.[17,18] However, as studies of unnatural vector-parasite associations lasted longer than those of natural combinations (mean (±s.e.m): 35.9 ±7.1 versus 15.1 ±3.3 days respectively), and study length influences virulence detection, no firm conclusions about the role of co-evolutionary history were possible.

More realistic estimates of malaria parasite virulence in mosquitoes would be obtained from direct observations in the field, where indirect costs of infection such as susceptibility to predation and anti-vector behaviour could be incorporated. However, these data are difficult to obtain as it is unethical to carry out mark-recapture experiments on experimentally infected mosquitoes. Some indirect evidence that free-living mosquitoes pay a survival cost from infection comes from the observation that sporozoites are unusually rare in large-bodied mosquitoes,[19] an observation interpreted as proof that mosquitoes with high parasite loads (most likely to be large mosquitoes as they take the biggest blood meals) suffer high mortality. However, large-bodied mosquitoes also have more effective immune responses (as shown for melanisation),[20,21] a phenomenon that could also explain the reduced sporozoite prevalence in this group. In contrast, Lines et al[22] interpreted the linear relationship between mosquito sporozoite prevalence and age as indirect evidence that malaria parasites do not reduce the survival of their vectors under natural conditions. Thus few firm conclusions can be drawn about the ubiquity of *Plasmodium*-imposed survival reduction in the field.

Fecundity

In contrast to investigations of mosquito survival, evidence that *Plasmodium* reduces vector fecundity is clear. To our knowledge, every study that has looked for a detrimental effect of *Plasmodium* on mosquito fecundity has found it. These studies involved a range of parasite-vector associations observed under both laboratory[23-31] and field conditions.[18] Typically these reductions result in a drop of egg production (following an infected meal) of approximately 15-35%.[15,28,30-32] Not only do mosquitoes produce fewer eggs after taking an infectious blood meal, but whilst they are infected with oocysts, the eggs they produce from uninfected blood have lower vitellin content and hatch rates than those produced by uninfected mosquitoes.[25]

Genetic and Environmental Determinants of Plasmodium Virulence in Vectors

Plasmodium virulence in mosquitoes could only be expected to evolve towards an optimum if it had a parasite genetic basis. If, for example, the fitness consequences of vector-parasite interactions were wholly phenotypic and determined solely by the particular details of the (often variable) environmental setting, evolution towards a virulence optima would not occur. However, recent laboratory studies of the rodent malaria *P. chabaudi* indicate that parasite genotype is a determinant of the magnitude of infection-induced reduction in both mosquito survival and fecundity.[31-33] *Plasmodium* virulence also depends on vector genetics, with parasites causing varying levels of harm to mosquitoes of different genotypes.[28] Superimposed upon these genetic determinants of virulence is environmental variation, which can also influence the magnitude of mosquito fitness reduction. One study showed that the most deadly *Plasmodium* genotype to mosquitoes under nutrient stress become the most benign when conditions were ideal,[33] whilst another found that parasite genetic differences in virulence were maintained under environmental variation.[32] Thus environmental conditions may mediate parasite genetic determinants of virulence, but do not appear to diminish them entirely. To definitively predict whether GM mosquitoes would benefit by avoiding parasitism, there is a need to measure the average virulence of malaria parasites in a range of mosquito genotypes.

Mechanisms of Parasite Virulence in Mosquitoes: Will Refractory Genes Block Them?

Understanding how virulence in mosquitoes could evolve under a GM strategy, requires knowledge of the basis of the fitness costs described above, and contemplation of whether the factors that drive them are likely to change under a control measure which manipulates mosquito immunity. Also, to predict whether GM mosquitoes would escape the costs of virulence requires knowledge of the efficiency with which they block infection. Currently the two strongest candidate refractory genes are those triggering expression of the peptides SM1[9] and PLA2.[10] Respectively, SM1 and PLA2 have been shown to reduce oocyst infection rates by 50-60%, oocyst intensity by 81-87%, and substantially cut the proportion of hosts that sporozoite-carrying mosquitoes infect.[9,10] However, neither peptide generates a complete transmission-blocking response. New approaches designed to transgenically upregulate the natural immune responses of mosquitoes instead of introducing a novel refractory gene have produced even less successful results, with only oocyst number but not infection rate being reduced.[34] Thus GM mosquitoes are not consistently parasite-free, they just have a lower probability of getting infected and harbouring high parasite burdens.

Many hypotheses have been proposed to explain the proximate causes of *Plasmodium* virulence in mosquitoes, although few have been tested. First of all, virulence could be a product of the physical damage parasites are known to inflict as they pass through the mosquito midgut, resulting in host cell apoptosis.[35-37] If this damage is the main cause of virulence, refractory genes could substantially reduce pathology by reducing the number of parasites that pass through the midgut. Virulence could also be the product of competition between mosquito tissues and growing parasites for limited energetic resources; and if so, refractory mosquitoes would again escape the cost of virulence by having smaller parasite burdens. However, resource competition is an unlikely cause of virulence. Less than a third of the energetic content in a blood meal is used for oogenesis,[15,38] thus protein for egg production should not be limiting even in the presence of *Plasmodium*. More critically, under resource limitation, the more parasites infecting a mosquito, the fewer eggs they should produce. No such relationship has been found.[15,31,32] The picture for survival is less clear, with mortality increasing with oocyst burden in some studies[26,27,32,39] but not others.[33,40] A further energetic explanation for the virulence of *Plasmodium* in mosquitoes is that it is not the parasites themselves that are harmful, but that infected blood is of poor nutritional quality. However, there appears to be no straightforward relationship between blood infection status and the amount of protein mosquitoes derive from it.[15] Mosquitoes have been found to take smaller blood meals from infected hosts,[26] but the extent of fecundity reduction amongst infected mosquitoes has been linked to blood meal size in only one study,[31] and not others.[26,32] The extent to which *Plasmodium* virulence in mosquitoes can be attributed to variation in blood meal size and quality is thus unresolved.

An additional potential explanation for the virulence of *Plasmodium* is that they elicit a costly immune response in mosquitoes. Mosquitoes are capable of mounting a diverse array of immune responses when invaded by pathogens.[41,42] *Plasmodium* infection triggers transcriptional activation of at least 6 different immune markers in *An. gambiae*, particularly when parasites are invading the midgut and salivary glands.[43] The production of these immune molecules could be energetically costly, and divert resources away from maintenance and reproduction.[44] Moreover, costs could also accrue if immunopathological damage occurs, as it almost always does in mammalian disease.[45] Some mosquitoes kill oocysts by melanotic encapsulation,[42] and one experimental study has shown that mosquitoes with the strongest encapsulation response have the highest mortality.[46] Furthermore, mosquitoes selected to be refractory to *Plasmodium* (through encapsulation of parasites) have poorer fitness than susceptible mosquitoes in the absence of infection.[47] If immune activation is costly (and of low efficacy) and the intensity of response is independent of parasite burden (as in ref. 44), it could explain why infected mosquitoes have reduced fitness. Unfortunately it has not yet been possible to test this hypothesis due to the difficulty of disentangling any costs resulting from direct parasite development from those generated by host immune activation. The prospect that *Plasmodium* virulence in mosquitoes is purely due to immune activation is problematic for the GM approach. It suggests that genetically refractory mosquitoes will receive no additional benefit from avoiding infection because their engineered phenotype (high immune responsiveness) is what causes *Plasmodium* to be virulent in the first place.

Evolution of Parasite Virulence in GM Mosquitoes

Selection Pressure on Transmission Selects for Higher Parasite Virulence: The Trade-Off Model

In recent years evolutionary biologists have devoted much attention to exploration of the conditions under which inflicting virulence may increase the transmission success and thus fitness of parasites.[48-52] The most dominant paradigm for virulence evolution is the trade-off model. The trade-off arises from the parasite's drive to maximize its multiplication rate in order to increase the number of propagules it can transmit without causing host death (and thus the truncation of transmission).[53] Critical assumptions of this model are that there is a genetic correlation between a parasite's multiplication rate and transmission success, and between multiplication rate and the risk of host death.[49,50,53] These predictions remain untested for the vast majority of infectious diseases, and the generality of the trade-off model remains contentious.[54]

However, there is now ample evidence to suggest it does hold for *Plasmodium* in its vertebrate host.[53] Detailed studies of both rodent and avian malaria parasites indicate that the number of transmission stages they produce in vertebrate blood increases with the rate of asexual parasite replication, and that asexual densities are correlated with host morbidity[53] and/or mortality.[53,55] Field data from the human malaria parasite, *P. falciparum*, are also strongly supportive.[53]

Do the Assumptions of the Virulence Trade-Off Model Apply in Malaria-Infected Mosquitoes?

In order to use the trade-off model to predict how parasite virulence in mosquitoes could evolve under GM pressure, it is necessary to evaluate if the trade-off appropriately describes the interaction between *Plasmodium* and their vectors. The first assumption of the trade-off model, that parasite multiplication rate is correlated with transmission potential, may not hold in mosquitoes. Within mosquitoes, *Plasmodium* multiplies as sporozoites that grow in oocysts before being released into the haemolymph and invading the salivary glands.[13] Sporozoites are injected into humans when infected mosquitoes blood feed. The number of sporozoites produced by individual oocysts of the human malaria *P. falciparum* ranges from 1000-4500,[56] with 10 ->130,000 ending up in the salivary glands of common African malaria vectors (geometric mean <1000).[13] Despite the fact that most oocyst infections generate thousands of sporozoites, only a small number (<25) are transmitted to a vertebrate host on biting.[13] In vitro, there appears to be no relationship between sporozoite load in salivary glands and the number that are transmitted upon biting.[57] However, in vivo studies give some evidence that the number of sporozoites transmitted during feeding increases the likelihood of infection.[58] Thus mosquitoes that inject high numbers, possibly as a result of having had a high replication rate with oocysts, may have a transmission advantage. A further complexity is that mosquitoes are frequently infected with several genetically distinct parasite genotypes; it seems likely that the frequency of individual genotypes within the sporozoite population will determine genotype fitness, as well as or in addition to numbers per se. Further research into the factors that mediate the transmission rates of individual genotypes to the vertebrate host, and particularly the effects of genotype multiplication rates in mosquitoes, is required. One transmission advantage of high sporozoite replication is that, at least in rodent models, they are associated with high per-sporozoite infectiousness.[58] Little is known about the cause of this density-dependent infectiousness, or whether it is a general phenomena of all malaria parasites. This phenomenon represents the strongest evidence for a positive association between *Plasmodium* replication in mosquitoes and transmission success as required by the trade-off model.

As for the second requirement that parasite replication increases vector mortality, evidence is mixed. Mosquito mortality increases with oocyst density in some[26,27,32,39] but not all studies.[33,40] To our knowledge, there has been no investigation of the relationship between sporozoite load in mosquitoes and mortality. Mosquitoes infected with sporozoites probe hosts more often,[59] are more persistent in returning to hosts after being disturbed,[60] and bite more people[61] than their uninfected counterparts. These behaviours could enhance parasite transmission success by increasing encounter rate with susceptible hosts, but may also increase mosquito exposure to anti-vector behaviour and thus their risk of death[62,63] (which may explain the observed increase in feeding-associated mortality in sporozoite infected mosquitoes).[64] Thus parasite-associated changes in feeding behaviour could generate the negative correlation between parasite replication and vector mortality required to substantiate application of the trade-off model of virulence evolution.

Further investigation of links between *Plasmodium* virulence in mosquitoes and transmission back to vertebrates is required before the relevance of the trade-off model can be evaluated. However, this preliminary discussion gives some evidence that the critical assumptions of this model may be met, and thus that virulence in mosquitoes could evolve under a control strategy in the same direction as predicted from the trade-off for *Plasmodium* in humans (e.g., ref. 65).

Parasite Virulence Evolution in Response to Manipulation of Host Immunity

Many disease control measures aim to enhance host resistance, for instance by vaccination or, in the case of agricultural animals, by selective breeding. Enhanced resistance will have a substantial impact on pathogen fitness. One evolutionary response which could occur is selection for variants that are able to evade these responses. The existence of *Plasmodium* strains which might evade mosquito immune recognition systems or be more resistant to the effectors produced by GM refractory genes is an important consideration, but beyond the scope of this chapter. Here, we discuss another possibility: in some cases, enhanced resistance can be expected to select for more virulent (aggressive) parasites.

Gandon et al[66] have recently made the following argument in the context of immunisation against human malaria. If more virulent *Plasmodium* parasites are removed by natural selection because they cause excessive host mortality and thus reduce their own fitness, keeping hosts alive with vaccines will allow more virulent strains to circulate in a population. This argument assumes that in the absence of host death, more virulent parasites transmit at higher rates, so that host death is the selective factors which stops the evolution of more virulent strains.

There is nothing in the mathematics of this argument which restricts it to vaccination. The key issue is the impact of resistance on the relative fitness of virulent and avirulent pathogens—however that resistance is generated. The argument thus applies to genetic resistance,[66,67] and here we ask how GM refractoriness in mosquitoes might impose selection for virulence evolution in *Plasmodium*. A key issue is the mode of action of the refractoriness.

Gandon et al[65] argued that there are four distinct types of disease resistance which are relevant to virulence evolution: (1) anti-infection, which reduces the probability of a host being infected, (2) anti-growth, which reduces the growth rate of malaria parasites within a host, (3) transmission-blocking, which reduces the infectiousness of an infection, and (4) anti-toxin resistance, which reduces the virulence of infection with no impact on the kinetics of infection. If resistance is sterilising (reduces transmission to zero), pathogen evolution cannot occur. However, if resistance is leaky, so that transmission is occurring, virulence evolution can occur. How it does so depends on which types of resistance are involved. Anti-growth and anti-toxin favour more virulent parasites strains, because they reduce host death and hence the fitness costs of virulence. In contrast, infection- and transmission-blocking resistance will not increase virulence. When these types of resistance are effective, the hosts are evolutionarily dead-ends; when they are not effective, the relative fitness of virulent and avirulent strains are unaltered, so that there is no selection in favour of greater resistance. In fact, transmission- or infection-blocking resistance can prompt mild reductions in virulence where within-host competition is affecting virulence evolution, if they reduce the number of competing genotypes.[65]

What Type of Host Resistance and Associated Parasite Evolution Could GM Mosquitoes Induce?

In essence, the GM approach is analogous to a vaccination programme for mosquitoes: an intervention aimed at cutting transmission by boosting the immunity of the mosquito host so it has a reduced probability of infection. To predict whether malaria parasites will evolve greater virulence in a population of GM mosquitoes requires identification of the mode of action of refractory genes. It is clear that current genetic constructs for refractoriness (e.g., AM1 and PLA2) cut infection, but we do not yet fully understand how they do this. It could be that they reduce the fertility of gametes or their ability to traverse the midgut, which would constitute anti-infection resistance, or they could lower the replication rate of sporozoites inside oocysts, which is anti-growth resistance. If some of the refractoriness conferred by these constructs or any developed in the future is partly anti-growth resistance, then the models of Gandon et al[65] apply: GM mosquitoes will prompt the evolution of strains which are more virulent to wild type mosquitoes. This argument assumes that the reason *Plasmodium* is not already more virulent to mosquitoes is that by killing their vector at higher rates, more virulent parasites have reduced fitness. Hence, if refractory genes enable mosquitoes to survive infections better, virulent strains that would have had lower fitness in a wild type mosquito are able to circulate in a GM population.

Such virulence evolution would reinforce the efficacy of GM mosquitoes as a means of malaria control. Higher parasite virulence would mean that fewer wild type mosquitoes would survive long enough to transmit malaria, thus reducing malaria transmission rates from wild type mosquitoes (see Box 1). It is difficult to predict how strong this evolutionary reinforcement would be, or how long the evolution would take to be detectable. In general models, there are regions of parameter space where near instantaneous evolutionary change is possible[66] other models suggest change could be in the order of a few decades or longer.[65] The more effective the refractoriness genes are, and the higher the frequency of GM mosquitoes in a population, the stronger the selection for virulence increases will be, but there will also be fewer malaria-infected wild type mosquitoes in which transmission-reducing effects of the virulence evolution accrue. Detailed models, parameterised for particular populations, will be needed to estimate the extent to which GM-prompted virulence evolution could contribute to reducing malaria transmission over and above that due to the immediate epidemiological effects of higher frequencies of refractory mosquitoes.

GM-induced refractoriness which is infection- or transmission-blocking will not prompt virulence evolution,[65] and would probably weaken any selection for virulence increases imposed by anti-growth or anti-toxin refractoriness. However, a further complexity would arise if, in genetically diverse malaria infections within mosquitoes, virulence schedules of individual genotypes effect the fitness of other genotypes. Such interactions seem likely: a clone highly virulent to mosquitoes is likely to reduce the longevity of any less virulent infections in the same mosquito. If they do occur, then infection-blocking refractoriness will reduce the genetic diversity of infections in mosquitoes, thus reducing in-host competition and selecting for parasite strains which are less virulent to mosquitoes. In principle, this evolution could lead to infected mosquitoes living for longer, and hence more malaria transmission. This would erode the efficacy of malaria control via GM mosquitoes, although the analogous effect in the vaccine case was relatively weak[65] (albeit with only a very limited range of competitive interactions modelled). Again, without parameterised models it is difficult to know how the epidemiological and evolutionary effects of refractoriness would combine to determine overall malaria transmission rates.

Could Plasmodium Virulence Evolution in Mosquitoes Impact Human Health?

The above arguments suggest that virulence evolution in mosquitoes is unlikely to reduce the efficacy of a GM control strategy, and could possibly enhance it. Does this mean *Plasmodium* virulence evolution in mosquitoes holds no detrimental consequences for public health? One disturbing possibility is that selection towards increased *Plasmodium* virulence in mosquitoes imposed by GM control would also give rise to parasite strains that are more virulent to humans. For example, the enhanced but incomplete immunity of GM mosquitoes could create selection for increased infectiousness, possibly by producing higher parasite burdens in humans in order to deliver a sufficiently large number of gametocytes in a blood meal to ensure some survive through the more intense immune response of a GM mosquito. In general, the ability of malaria-infected hosts to infect mosquitoes increases with their gametocyte density,[53] although high gametocytes densities do not always result in high mosquito infection rates[68,69] and other factors may play important roles.[70] Evolution towards increased gametocyte density would only be expected to occur if genetically-engineered refractoriness could be overcome simply by swamping the mosquito immune system with greater numbers of parasites, a hypothesis that has not been tested.

Box 1. Basic reproductive ratio with genetically modified mosquitoes

The mathematical study of malaria epidemiology began with Ross (1911)[78] who showed that there is a critical density of mosquitoes below which transmission can be eliminated. This result is of paramount importance because it proves that the burden of malaria can be reduced by fighting mosquitoes. A few decades later, the further mathematical developments of Macdonald[79,80] led him to devise a transmission measure known the basic reproductive ratio, R_0, which is the expected number of secondary malaria infections resulting from a single case breaking out in an uninfected host population. This famous quantity is particularly useful because it summarises the whole parasite life cycle and can be used to evaluate the degree by which mosquito and parasite parameters must be altered to cause eradication (eradication is feasible when R_0 can be reduced below 1). In particular this quantity helped Macdonald to identify the most vulnerable element in the malaria cycle: the survivorship of adult female *Anopheles*. Macdonald suggested that 'the worst conditions known in Africa could therefore be overcome by an increase in the daily mortality of the vector from about 5% to about 45%.[79] This provided the rationale for campaigns of eradication based on the use of DDT. This strategy did not yield world-wide eradication, but it did help reduce prevalence temporarily and regionally (malaria has been eradicated from several countries).

In a simple model (here we assume homogeneous mixing and fixed densities of humans and mosquitoes), the basic reproductive rate of malaria under a GM strategy can be defined as follows:

$$R_0 = \frac{a^2 m}{d_H + v_H}\left((1-p)\frac{b_H b_V e^{-(d_V + v_V)T}}{d_V + v_V} + p\frac{b'_H b'_V e^{-(d'_V + v'_V)T'}}{d'_V + v'_V} \right)$$

where:

a: biting rate of mosquitoes on humans

m: total number of mosquitoes per human host

p: proportion of mosquitoes that are genetically modified (GM) to be refractory

b_H, b'_H: transmission rate from infected humans to uninfected wildtype and GM mosquitoes

b_V, b'_V: transmission rate from infectious wildtype and GM mosquitoes back to uninfected humans

d_H, d_V, d'_V: intrinsic death rate of humans, wildtype and GM mosquitoes respectively

v_H, v_V, v'_V: malaria virulence in humans, wildtype and GM mosquitoes respectively

T, T': development time of the parasite in wildtype and GM mosquitoes

Note that, for the sake of simplicity, intrinsic death rates (d_V, d'_V) and virulence (v_V, v'_V) are assumed to be constant throughout the development of malaria in infected mosquitoes (no differences between infectious and infected but not yet infectious mosquitoes).

The above expression can be used to quantify the effects of various properties of GM mosquitoes (denoted with primes) on R_0 and, therefore, can be used to predict the potential success of different GM strategies. It illustrates that maximal reduction of R_0 requires both (1) increasing the relative abundance (*p*) of GM mosquitoes and (2) reducing the vectorial capacity of GM mosquitoes

$$\left(\frac{b'_H b'_V e^{-(d'_V + v'_V)T'}}{d'_V + v'_V} \right).$$

Currently, attempts to reduce vectorial capacity have been based on increasing the immune response of GM mosquitoes to malaria. Additionally, we note, that if a sufficiently high number of GM mosquitoes could be introduced into a population (p), another potential means to reduce vectorial capacity would be to make them extra sensitive to parasitism so that they died before becoming infectious; or simply increasing their instrinsic mortality rate so that none of them survived the parasite's development period. In other words, the vector population would be replaced not by immune mosquitoes, but by ones unable to support parasite development because of poor survival.

Box continued on next page

Box 1.   Continued

Increasing the intrinsic mortality of GM mosquitoes (d'_V) is analogous to the recommendation of Macdonald[79] but increasing the virulence of malaria on GM mosquitoes would be an original approach. Neither of these longevity-targeting strategies would work unless GM mosquitoes were sufficiently abundant to yield immediate epidemiological effects; as in small numbers their survival disadvantage would only increase the rate at which they were eliminated from populations.

The relative abundance (p) of GM mosquitoes will of course depend on the "release" effort, but also on the life-history traits of both wild and GM mosquitoes. In particular, note that the above recommendations based on reducing transmission rates and/or vector survival may conflict with the goal of reaching a large density of GM mosquitoes. For example, if the resistant GM mosquitoes (b'_H, b'_V is low) carry a large cost (for example on fecundity) they will be poor competitors against wild mosquitoes and will never reach high densities (i.e., p will remain low). Similarly, if GM mosquitoes are very susceptible to dying from malaria (v'_V is high) and if the prevalence of malaria is high, they will never reach a high density. Analysis of the consequences of modifying these life-history traits of GM mosquitoes requires consideration of these epidemiological feed-backs.[76] Malaria evolution and adaptation to GM mosquitoes may also introduce other unexpected feed-backs on transmission, as discussed here and elsewhere.[2]

More generally, should we expect selection for increased virulence in one host to generate a correlated response in another? This depends on the genetic correlations between the factors that underlie virulence in both hosts (e.g., high parasite burdens).[71] Studies of schistosomes indicate that strains most deadly to their intermediate snail host (characterized by low replication rate) also elicit the most harm in their vertebrate host (where their replication rate is high),[72] and moreover that selection for increased parasite replication in snail host generates a correlated drop in virulence in both hosts.[73] Thus, virulence in the two hosts is apparently positively and genetically correlated. If this correlation is not eroded by contrasting selection pressures in different hosts, then selection for increased schistosome virulence in snails would result in increased disease severity in their vertebrate hosts.

Could this happen in malaria? Studies of the rodent malaria *P. chabaudi* suggest that disease severity in mice is positively correlated with mosquito infection rate,[32,74] but was not associated with mosquito survival.[32] This study, the only one of malaria virulence in vertebrates and vectors of which we are aware, suggests there is no link between malaria virulence in vertebrates and vector mortality. Thus, with the caveat that all available knowledge comes from just one study of a rodent model, it seems unlikely that any selection for increased *Plasmodium* virulence in mosquitoes imposed by GM refractoriness will generate a correlated increase in parasite virulence in humans. More such studies should be undertaken if deployment of GM mosquitoes becomes a serious possibility and precaution should be the rule.

Conclusions

As malaria parasites are virulent in their mosquito vectors, the fitness cost to mosquitoes of refractoriness imposed by transgenesis could in theory be partly eroded by the benefit of not becoming infected. Unfortunately, the currently best candidate molecules for engineering refractoriness, SM1 and PLA2, do not completely block transmission but instead reduce parasite burden. As reviewed above, malaria-induced reductions in mosquito fecundity are independent of parasite burden, so that mosquitoes with one parasite have the same fitness loss as those that become heavily infected. This means that unmodified mosquitoes carrying 100 oocysts may have equal fitness to those carrying refractory genes that only permit the development of only 1-2 oocysts; and thus that there is no substantial benefit to the mosquito of being more resistant. Malaria-induced reductions in mosquito survival may be linked to parasite burdens, so SM1 and PLA2 could enhance that component of mosquito fitness, although the survival costs of malaria infection have proven hard to detect and so seem likely to be relatively small.

Could reductions in *Plasmodium* virulence to mosquitoes of the magnitude reviewed here ever outweigh the substantial fitness costs of becoming refractory?[3-5,47] A sceptic would argue that the cost of refractoriness is always greater than that imposed by *Plasmodium*, because otherwise mosquitoes would already have evolved immunity towards malaria parasites. This may be an overly simplistic view: co-evolutionary dynamics generally favour the partner with the short generation time (in this case, parasites), and can maintain low virulence even when the costs of immunity are not high.[75,76] Moreover, the efficiency with which natural selection could act to eliminate *Plasmodium* susceptibility in free-living vectors may be low given that only a small proportion (1-2%)[13] are ever exposed to infection. Yet under a GM strategy all mosquitoes in the population would express a refractory gene and thus all would pay for any fitness cost it carried. This population-level effect could generate more intense and efficient selection for mosquitoes to lose immunity than *Plasmodium* generates to develop it. Quantitative analysis based on field-derived estimates of mosquito fitness is required to assess how the costs of virulence and immunity weigh up before their relevance to GM control can be assessed.

The direction under which malaria parasite virulence in mosquitoes will evolve under a GM strategy is unknown, and will likely depend upon the type of refractory genes selected for use and their impact on mosquito fitness. Ideally the widespread application of transgenesis in vector populations would mimic the impacts of vaccination with an imperfect anti-parasite growth vaccine; a strategy which the simplest models predict will increase the deadliness of parasites in nonvaccinated hosts.[65] This evolutionary outcome could boost the success of a GM campaign by simultaneously increasing the proportion of refractory vectors whilst decreasing the survival of susceptibles. Further experimental and theoretical research into the evolutionary response of *Plasmodium* in the face of genetically-engineered immunity is required to evaluate whether this outcome is possible.

Acknowledgements

Empirical and theoretical work discussed here was supported by the Wellcome Trust, the Royal Society and the CNRS. We thank members of the Read group at the University of Edinburgh and the Public Health Entomology Unit at the IHRDC for discussion.

References

1. Scott TW, Takken W, Knols BG et al. The ecology of genetically modified mosquitoes. Science 2002; 298(5591):117-119.
2. Boëte C, Koella JC. Evolutionary ideas about genetically manipulated mosquitoes and malaria control. Trends Parasitol 2003; 19(1):32-38.
3. Catteruccia F, Godfray HC, Crisanti A. Impact of genetic manipulation on the fitness of Anopheles stephensi mosquitoes. Science 2003; 299(5610):1225-1227.
4. Moreira LA, Wang J, Collins FH et al. Fitness of Anopheline mosquitoes expressing transgenes that inhibit Plasmodium development. Genetics 2004; 166:1337-1341.
5. Irvin N, Hoddle MS, O'Brochta DA et al. Assessing fitness costs for transgenic Aedes aegypti expressing the GFP marker and transposase genes. Proc Natl Acad Sci USA 2004; 101(3):891-896.
6. Boëte C, Koella JC. A theoretical approach to predicting the success of genetic manipulation of malaria mosquitoes in malaria control. Malar J 2002; 1(1):3.
7. Riehle MA, Srinivasan P, Moreira CK et al. Towards genetic manipulation of wild mosquito populations to combat malaria: advances and challenges. J Exp Biol 2003; 206:3809-3816.
8. Gould F, Schliekelman P. Population genetics of autocidal control and strain replacement. Annu Rev Entomol 2004; 43:193-217.
9. Ito J, Ghosh A, Moreira LA et al. Transgenic anopheline mosquitoes impaired in transmission of a malaria parasite. Nature 2002; 417(6887):452-455.
10. Moreira LA, Ito J, Ghosh A et al. Bee venom phospholipase inhibits malaria parasite development in transgenic mosquitoes. J Biol Chem 2002; 277(43):40839-40843.
11. Woolhouse MEJ, Webster JP, Domingo E et al. Biological and biomedical implications of the co-evolution of pathogens and their hosts. Nature Genetics 2002; 32:569-577.
12. Elliot SL, Adler FR, Sabelis MW. How virulent should a parasite be to its vector? Ecology 2003; 84(10):2568-2574.
13. Beier JC. Malaria parasite development in mosquitoes. Annu Rev Entomol 1998; 43:519-543.

14. Anderson RM, May RM. Infectious diseases of humans: Dynamics and control. New York: Oxford University Press; 1991.
15. Hurd H. Manipulation of medically important insect vectors by their parasites. Annu Rev Entomol 2003; 48:141-161.
16. Ferguson HM, Read AF. Why is the effect of malaria parasites on mosquito survival still unresolved? Trends Parasitol 2002; 18:256-261.
17. Chege GMM, Beier JC. Effect of Plasmodium falciparum on the survival of naturally infected Afrotropical Anopheles (Diptera: Culicidae). J Med Entomol 1990; 27:454-458.
18. Hogg JC, Hurd H. The effects of natural Plasmodium falciparum infection on the fecundity and mortality of Anopheles gambiae s.l. in north east Tanzania. Parasitology 1997; 114:325-331.
19. Lyimo EO, Koella JC. Relationship between body size of adult Anopheles gambiae sl and infection with the malaria parasite Plasmodium falciparum. Parasitology 1992; 104(Pt2):233-237.
20. Schwartz A, Koella JC. Melanization of Plasmodium falciparum and c-25 sephadex beads by field-caught Anopheles gambiae (Diptera: Culicidae) from southern Tanzania. J Med Entomol 2002; 39:84-88.
21. Boëte C, Paul R, Koella J. Reduced efficacy of the immune melanization response in mosquitoes infected by malaria parasites. Parasitology 2002; 125:93-98.
22. Lines J, Wilkes T, Lyimo E. Human malaria infectiousness measured by age-specific sporozoite rates in Anopheles gambiae in Tanzania. Parasitology 1991; 102:167-177.
23. Carwardine SL, Hurd H. Effects of Plasmodium yoelii nigeriensis infection on Anopheles stephensi egg development and resorption. Med Vet Entomol 1997; 11(3):265-269.
24. Hopwood J, Ahmed A, Polwart A et al. Malaria-induced apoptosis in mosquito ovaries: a mechanism to control vector egg production. J Exptl Biol 2001; 204:2773-2780.
25. Ahmed AM, Maingon RD, Taylor PJ et al. The effects of infection with Plasmodium yoelii nigeriensis on the reproductive fitness of the mosquito Anopheles gambiae. Invert Reprod & Dev 1999; 36(1-3):217-222.
26. Hogg JC, Hurd H. Malaria-induced reduction of fecundity during the first gonotrophic cycle of Anopheles stephensi mosquitoes. Med Vet Entomol 1995; 9:176-180.
27. Hogg JC, Hurd H. Plasmodium yoelii nigeriensis: the effect of high and low intensity of infection upon the egg production and bloodmeal size of Anopheles stephensi during three gonotrophic cycles. Parasitology 1995; 111:555-562.
28. Hacker CS. The differential effect of Plasmodium gallinaceum on the fecundity of several strains of Aedes aegypti. J Invert Pathol 1971; 18:373-377.
29. Hacker CS, Kilama WL. The relationship between Plasmodium gallinaceum density and the fecundity of Aedes aegypti. J Invert Pathol 1974; 23:101-105.
30. Freier JE, Friedman S. Effect of host infection with Plasmodium gallinaceum on the reproductive capacity of Aedes aegypti. J Inverteb Pathol 1976; 28:161-166.
31. Ferguson H, Rivero A, Read A. The influence of malaria parasite genetic diversity and anaemia on mosquito feeding and fecundity. Parasitology 2003; 127:9-19.
32. Ferguson H, Mackinnon M, Chan B et al. Mosquito mortality and the evolution of malaria virulence. Evolution 2003; 57(12):2792-2804.
33. Ferguson HM, Read AF. Genetic and environmental determinants of malaria parasite virulence in mosquitoes. Proc R Soc Lond B 2002; 269(1497):1217-1224.
34. Kim W, Koo H, Richman AM et al. Ectopic expression of a cecropin transgene in the human malaria vector mosquito Anopheles gambiae (Diptera: Culicidae): effects on susceptibility to Plasmodium. J Med Ent 2004; 41(3):447-455.
35. Han YS, Thompson J, Kafatos FC et al. Molecular interactions between Anopheles stephensi midgut cells and Plasmodium berghei: the time bomb theory of ookinete invasion of mosquitoes. EMBO J 2000; 19:6030-6040.
36. Zieler H, Dvorak JA. Invasion in vitro of mosquito midgut cells by the malaria parasite proceeds by a conserved mechanism and results in the death of the invaded midgut cells. Proc Natl Acad Sci USA 2000; 97:11516-11521.
37. Ramasamy MS, Kulaksekera R, Wanniarachchi IC et al. Interactions of human malaria parasites Plasmodium vivax and P. falciparum, with the midgut of Anopheles mosquitoes. Med Vet Entomol 1997; 11:290-296.
38. Briegel H. Fecundity, metabolism, and body size in Anopheles (Diptera, Culicidae), vectors of malaria. J Med Entomol 1990; 27(5):839-850.
39. Klein TA, Harrison BA, Grove JS et al. Correlation of survival rates of Anopheles dirus (Diptera: Culicidae) with different infection densities of Plasmodium cynomolgi. Bull WHO 1986; 64(6):901-907.

40. Gamage-Mendis AC, Rajaruna J, Weerasinghe S et al. Infectivity of Plasmodium vivax and Plasmodium falciparum to Anopheles tessellatus; relationship between oocyst and sporozoite development. Trans R Soc Trop Med Hyg 1993; 87:3-6.

41. Barillas-Mury C, Wizel B, Han YS. Mosquito immune responses and malaria transmission: lessons from insect model systems and implications for vertebrate innate immunity and vaccine development. Insect Biochem. Molec Biol 2000; 30(6):429-442.

42. Carton Y, Nappi AJ, Poirie M. Genetics of anti-parasite resistance in invertebrates. Develop. Comparative Immunol 2005; 29:9-32.

43. Dimopoulos G. Insect immunity and its implication in mosquito-malaria interactions. Cellular Microbiol 2003; 5(1):3-14.

44. Ahmed A, Baggott S, Maingon R et al. The costs of mounting an immune response are reflected in the reproductive fitness of the mosquito Anopheles gambiae. Oikos 2002; 97:371-377.

45. Graham AL, Allen JE, Read AF. Evolutionary causes and consequences of immunopathology. Annu Rev Ecol Syst 2005; 36:373-397.

46. Boëte C, Paul RE, Koella JC. Direct and indirect immunosuppression by a malaria parasite in its mosquito vector. Proc R Soc Lond B Biol Sci 2004; 271:1611-1615.

47. Yan G, Severson DW, Christensen BM. Costs and benefits of mosquito refractoriness to malaria parasites: Implications for genetic variability of mosquitoes and genetic control of malaria. Evolution 1997; 51(2):441-450.

48. Anderson RM. Population dynamics of infectious diseases: theory and applications. London: Chapman and Hall; 1982.

49. Frank SA. Models of parasite virulence. Q Rev Biol 1996; 71(1):37-78.

50. Bull JJ. Perspective—Virulence. Evolution 1994; 48(5):1423-1437.

51. Ebert D, Herre EA. The evolution of parasitic diseases. Parasitol Today 1996; 12(3):96-101.

52. Gandon S. Evolution of multihost parasites. Evolution 2004; 58(3):455-469.

53. Mackinnon MJ, Read AF. Virulence in malaria: an evolutionary viewpoint. Phil Trans R Soc Lond B 2004; 359:965-986.

54. Ebert D, Bull JJ. Challenging the trade-off model for the evolution of virulence: is virulence management feasible? Trends Microbiol 2003; 11(1):15-20.

55. Paul REL, Lafond T, Muller-Graf CDM et al. Experimental evaluation of the relationship between lethal or nonlethal virulence and transmission success in malaria parasite infections. BMC Evolution Biol 2004; 4:30.

56. Rosenberg R, Rungsiwongse J. The number of sporozoites produced by individual malaria oocysts. Am. J Trop Med Hyg 1991; 45(5):574-577.

57. Beier JC, Onyango FK, Koros JK et al. Quantitation of malaria sporozoites transmitted in vitro during salivation by wild Afrotropical Anopheles. Med Vet Entomol 1991; 5:71-79.

58. Pumpuni CB, Mendis C, Beier JC. Plasmodium yoelii sporozoite infectivity varies as a function of sporozoite loads in Anopheles stephensii mosquitoes. J Parasitol 1997; 83(4):652-655.

59. Wekesa JW, Copeland RS, Mwangi RW. Effect of Plasmodium falciparum on blood feeding behaviour of naturally infected Anopheles mosquitoes in western Kenya. Am. J Trop Med Hyg 1992; 47:484-488.

60. Anderson RA, Koella JC, Hurd H. The effect of Plasmodium yoelii nigeriensis infection on the feeding persistence of Anopheles stephensi Liston throughout the sporogonic cycle. Proc R Soc Lond B 1999; 266(1430):1729-1733.

61. Koella JC, Sorensen FL, Anderson RA. The malaria parasite, Plasmodium falciparum, increases the frequency of multiple feeding of its mosquito vector, Anopheles gambiae. Proc R Soc Lond B 1998; 265:763-768.

62. Schwartz A, Koella JC. Trade-offs, conflicts of interest and manipulation in Plasmodium-mosquito interactions. Trends Parasitology 2001; 17(4):189-194.

63. Anderson RA, Roitberg BD. Modelling trade-offs between mortality and fitness associated with persistent blood feeding by mosquitoes. Ecology letters 1999; 2:98-105.

64. Anderson RA, Knols BG, Koella JC. Plasmodium falciparum sporozoites increase feeding-associated mortality of their mosquito hosts Anopheles gambiae s.l. Parasitology 2000; 129:329-333.

65. Gandon S, Mackinnon M, Nee S et al. Imperfect vaccines and the evolution of virulence. Nature 2001; 414(6865):751-756.

66. Gandon S, Mackinnon M, Nee S et al. Imperfect vaccination: some epidemiological and evolutionary consequences. Proc R Soc Lond B Biol Sci 2003; 270:1129-1136.

67. Read AF, Gandon S, Nee S et al. The evolution of pathogen virulence in response to animal and public health interventions. In: Dronamraj K, ed. Infectious Disease and Host-Pathogen Evolution. Cambridge: Cambridge University Press 2004.

68. Jefferey GM, Eyles DE. Infectivity to mosquitoes of Plasmodium falciparum as related to gameto-cyte density and duration of infection. Am J Trop Med Hyg 1955; 4:781-789.
69. Boudin C, Olivier M, Molez JF et al. High human malaria infectivity to laboratory-bred Anoph-eles gambiae in a village in Burkina Faso. Am J Trop Med Hyg 1993; 48(5):700-706.
70. Paul RE, Diallo M, Brey PT. Mosquitoes and transmission of malaria parasites—not just vectors. Malar J 2004; 3:39.
71. Gandon S, Agnew P, Michalakis Y. Coevolution between parasite virulence and host life-history traits. Am Nat 2002; 160(3):374-388.
72. Davies CM, Webster JP, Woolhouse MEJ. Trade-offs in the evolution of virulence in an indirectly transmitted macroparasite. Proc R Soc Lond B 2001; 268(1464):251-257.
73. Gower CM, Webster JP. Fitness of indirectly transmitted pathogens: restraint and constraint. Evo-lution 2004; 58(6):1178-1184.
74. Mackinnon MJ, Read AF. Genetic relationships between parasite virulence and transmission in the rodent malaria Plasmodium chabaudi. Evolution 1999; 53(3):689-703.
75. Van Baalen M. Coevolution of recovery ability and virulence. Proc R Soc Lond B 1998; 265:317-325.
76. Koella JC, Boëte C. A model for the coevolution of immunity and immune evasion in vector-borne diseases with implications for the epidemiology of malaria. Am Nat 2003; 161(5):698-707.
77. MetaWin: Statistical Software for Meta-Analysis [computer program]. Version 2.0. Sunderland, Massachussetts: Sinauer Associates; 2000.
78. Ross R. The prevention of malaria. London: Murray; 1911.
79. Macdonald G. The epidemiological basis of malaria control. Bull WHO 1956; 15:613-626.
80. Macdonald G. The Epidemiology and Control of Malaria. London: Oxford University Press; 1957.

CHAPTER 10

Thinking Transgenic Vectors in a Population Context:
Some Expectations and Many Open-Questions

Christine Chevillon,* Richard E. Paul, Thierry de Meeûs
and François Renaud

Abstract

The present chapter tries to place questions regarding the eventual release of transgenic *Plasmodium*-resistant mosquitoes within an overall population context. This means a context that is not limited to selection outcomes regarding the resistance/susceptibility of the *Anopheles* targeted by transgenesis, but that also recovers selection outcomes regarding other traits, demographic outcomes (via the possible number of transgenic mosquitoes to release), and biodiversity outcomes regarding the vectors and parasites that cocirculate in the chosen release locality. By considering all these outcomes together, we highlight missing biological data necessary for any correct quantitative evaluation of the probability of success of a transgenic release. Qualitative evaluations are nonetheless possible to perform: they suggest that there is a very weak probability for released transgenic mosquitoes to actually succeed in modifying malaria transmission. However, the main interest of the present discussion does not concern these qualitative conclusions. Instead, it highlights the necessary biological knowledge of malaria for a correct evaluation of the fate and consequences of eventual releases of transgenic vectors, and more generally of the evolutionary possibilities and constraints of any change in transmission characteristics. As such, we hope that the present discussion underlines the necessity to address new fundamental questions regarding malaria biology in order to actually capture the mechanics regulating the evolutionary dynamics of *Plasmodium* burdens and associated pathologies.

From the outset, it is assumed that mosquito transgenesis, inducing *Anopheles* resistance to *Plasmodium* infection, becomes sufficiently "routine" such that the major issue would concern the choice of the candidate to release in order to maximize public health benefit. From then, we explore the extent to which concepts of population genetics and evolutionary biology may help in evaluating, or even optimizing, the chance of a successful transgenic release strategy. One approach would be to develop explicit mathematical models targeting the epidemiological consequences of the vector evolution toward parasite-resistance. This is fruitful[1] but requires *a priori* assumptions, concerning the estimates taken by several population genetics parameters, whose biological pertinence may be difficult to evaluate. Thus, we chose an alternative

*Corresponding Author: Christine Chevillon—Team 'Evolution des Systèmes Symbiotiques',
Laboratoire GEMI (UMR IRD-CNRS 2724), Centre IRD, 911 avenue Agropolis BP 64501,
F 34 394 Montpellier Cedex 5, France. Email: christine.chevillon@mpl.ird.fr

Genetically Modified Mosquitoes for Malaria Control, edited by Christophe Boëte.
©2006 Landes Bioscience.

approach with the hope of strengthening transdisciplinary discussions regarding the practical constraints linking transgenic production to release. Therefore, we have tried (i) to clearly define the population genetics parameters that both determine the fate of transgenic mosquitoes in natural populations and are easy to measure in experimental set-ups; (ii) to make explicit the assumptions hidden behind the structure of population genetics models, and (iii) to confront assumptions with biological data. Finally, where knowledge is lacking, we have tried to underline the neglected biological questions that would provide invaluable information.

This framework is applied to three complementary issues that delimit the sections of the present chapter. The first section is aimed at identifying the early risks of transgene disappearance and the precautions to take in order to minimize these risks. The second section defines the parameters, accessible to experimental evaluation, that determine the long-term evolution of the released resistance transgene and their epidemiological consequences under idealized conditions. We have also tried to show how incorporation of increasing levels of biological reality may alter conclusions. This enables clear identification of the biological data necessary for a correct evaluation of the epidemiological consequences of any release of *Plasmodium*-resistant mosquitoes. Such requisite data will be confronted with the current knowledge on malaria biology in a third and final section, exposing several gaps between required and acquired knowledge. Overall, this discussion hopes to highlight, not only the reasons why evolutionary biologists are so skeptic about the public health benefits to expect from transgenic resistance release, but also the field and experimental studies that we must address in order to understand the mechanics regulating the evolutionary dynamics of *Plasmodium* burdens and associated pathologies.

Transgenic Naturalization: Considerations for Successful Invasion

It was claimed that *'first transgenics released must be sterile'*.[2] As a preliminary argument, let us wipe out a possible confusion between two mutually exclusive processes aimed at modifying the epidemiology of vector-borne diseases. One strategy was the release of sterile vector males (possibly resulting from transgenesis, see ref. 2 for review). These sterile males have been hoped to reduce vector demography through competition between sterile and wild males for wild females. For this strategy to have an impact, humans should produce and release, at each mosquito generation, a number of sterile males that matches that of the wild-type females seeking a mate, or even higher number if females can have several mates; *An. gambiae* females mate more than once,[3] and frequently occur in large numbers within populations. Thus, it is not surprising that this 'sterile male' strategy failed when applied to this species complex.[4]

Examining the potential of releasing *Plasmodium*-resistant mosquitoes radically changes the rationale. Indeed, a transgene that confers resistance to *Plasmodium* infection can be effective if and only if it is designed to be expressed by female mosquitoes blood-feeding on humans. This is because only female mosquitoes face infection and they do so while blood-feeding. Therefore, whenever one considers such a strategy to improve public health, he/she automatically considers releasing transgenic mosquitoes, of either sex, that will lay fecund transgenic female descents in nature. If modifying the genetic composition of a vector population and letting people being bitten by transgenic females raise ethical concerns,[2] it is noteworthy that these concerns are actually inherent to any success using a *Plasmodium*-resistance release.

But what does determine the probability of success of such a strategy? It depends on the relative fitness of transgenic mosquitoes and on that of the transgene itself. The fitness of transgenic relative to wild-type mosquitoes can be measured by their average difference in offspring number. The fitness of the transgene corresponds to the average number of copies generated per generation relative to the equivalent average computed for a standard nonselected ('neutral') gene. Equivalence between these fitnesses only occurs when transgene carriers share the same genetic background as wild-type mosquitoes. This condition characterizes what we call a 'naturalized transgene' (Fig. 1A). Thus, the process clearly discriminates a prenaturalization period, when the success of the transgenic strategy is almost independent of transgene fitness, and a post-naturalization period when success is tightly linked to transgene fitness. The case of

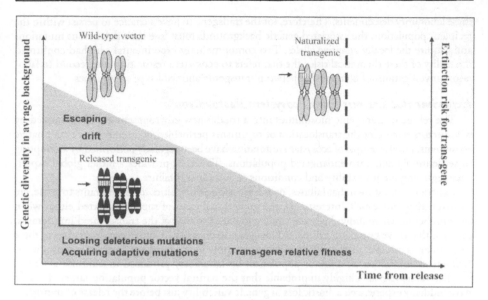

Figure 1A. The dangers faced by laboratory-selected genes when released in populations: intensities, causes and protection means. At release time, a *Plasmodium* resistance-gene (figured here as a locus pointed by an arrow) is necessarily borne by a laboratory genetic background. This laboratory genetic background is likely to be on average more inbred and to bear more homozygous deleterious mutations that the average wild-type one (see text). We draw here deleterious mutations as white loci, genetic backgrounds of low diversity as black chromosomes, of high diversity as grey chromosomes, intermediate diversity levels as chromosomes bearing black and grey motives, with in this latter case identify/difference in motives reflecting genetic identity/differences. Figure 1A pictures the differences regarding the genetic backgrounds of a wild-type vector and of a transgenic mosquito taken either at release time or at the end of the naturalization period (dotted line). The grey area illustrates the time-evolution of extinction risks for the transgene. The sources of extinction risk are indicated in bold characters with reference to their relative time of predominance. Please note that the highest risks occur very early after the release time and that the transgene fitness will only become important near the end of the naturalization period.

post-naturalization period will be discussed in the section "Evaluating the Chance of Success of Naturalized Resistance Genes: Formalization and Estimation of the Selective Balance Involved".

Drift, Inbreeding and Background Selection: The Three Major Risks in Prenaturalization

The extinction risks that the transgene faces before it achieves naturalization originate from three properties of any laboratory-strain release (Fig. 1A). First, the relative frequency of the released transgenic mosquitoes compared to wild-type ones will be relatively low. This not only opens the road for genetic drift to accidentally clear the recipient population from any transgene carrier, but also to do it rapidly and independently of any fitness consideration.

Even if enough mosquitoes are released to escape the immediate risk of drift, two other dangers are likely to arise because of the higher inbreeding expected among laboratory-released compared to wild-type mosquitoes. The released transgenic mosquitoes are descended from one or a few strains that have evolved for many generations under laboratory conditions. Accordingly, genetic drift and selection for adaptation to laboratory environments are expected to have (i) progressively increased the genetic divergence between laboratory-descendents and wild ancestors at many loci (including those that can be advantageous in the environment of release), (ii) progressively reduced the genetic diversity of laboratory descendents down through the generations, and (iii) fixed by chance a few deleterious mutations into the genetic background of these laboratory descendents. Therefore, for the transgene to have a chance to persist within the recipient population, the associated genetic backgrounds must lose their deleterious mutations and acquire the locally adaptive alleles. Two complimentary experimental axes had confirmed the reality of these theoretical risks: the first refers to ecosystem restoration, the second to laboratory investigations of competition between transgenic and wild-type mosquitoes.

Experimental Lessons from Ecosystem Restoration

The release of transgenic mosquitoes into a totally new environment already occupied by wild vectors resembles the translocation of organisms performed in attempts to restore native ecosystems. Local attempts of ecosystem restoration have been regularly performed by attempting to settle immigrants into endangered populations. This thus provides a rather global experience regarding the probability and conditions of success and failure.[5]

Experience has shown that failures, or at least strong difficulties for immigrants to settle, are the overriding outcome.[5] Interestingly, the exceptional cases of success occurred either when the number of native individuals was small compared to that of the transplanted immigrants, or when the native population displayed low genetic diversity.[5] Unfortunately, neither of these two scenarios can ever reasonably be expected to apply to the release of transgenic mosquitoes. Numerical dominance of a vector population by laboratory-produced mosquitoes is almost inconceivable. It is also highly improbable that the natural vector population targeted would have suddenly experienced a drastic loss in genetic variability just before the release of transgenic vectors.

By contrast, the difficulties recurrently encountered by immigrants in ecosystem restorations are very likely to apply to the release of transgenic mosquitoes. A first difficulty arises because the transplanted immigrants bear genotypes that confer adaptation to their ancestral but not to their new environments.[5,6] Given the large differences between laboratory-controlled and field environments, it is very likely that such mal-adaptation will also concern the laboratory-engineered mosquitoes at release time. A second difficulty arises because immigrants display too low a genetic diversity and too much inbreeding to get rid of their deleterious mutations within their new challenging environment.[5] Avoidance of inbreeding is a difficult goal to achieve in laboratory-reared strains, requiring either the application of laborious mating protocols or regular incorporation of individuals from foreign stocks. Therefore, this second difficulty is also very likely to apply to released transgenic mosquitoes, and indeed has been directly confirmed in laboratory population experiments.

Transgene Naturalization in Laboratory Experiments: The Actual Dimension of Inbreeding

Transgenesis is rarely successful at 100% so that transgenic strains are generally founded by a very few individuals. Therefore, the risk for a chance fixation of deleterious mutations looks even greater within transgenic than in standard laboratory strains. A recent study investigated this question by allowing transgenic and wild-type mosquitoes to compete within experimental populations maintained under laboratory conditions (to which all competitors were adapted).[7] Four transgenic lines of *Anopheles stephensi* were involved where transgenes encoded distinct fluorescent proteins. Genetic drift was avoided by seeding populations with a 50:50 mix of transgenic and nontransgenic mosquitoes. In all replicates and whatever the identity of the transgene, the outcome was disappearance of the transgene from experimental populations within a few generations. Making the wild-type mosquitoes as inbred as the transgenic ones was enough to considerably increase the number of generations during which the transgene persisted.[7]

A later study reinvestigated this issue by focusing on two transgenic constructs that make *Anopheles* mosquitoes resistant to *Plasmodium* infection.[8] A first transgenic construct encodes a tetramer of SM1 peptide. This peptide prevents vector infection by competitively binding to the *Anopheles* receptors that *Plasmodium* parasites use to infect the vector.[9] In this case, homogenization in genetic background among competing mosquitoes was sufficient to ensure transgene naturalization in all replicates originally seeded with half wild-type and half transgenic mosquitoes.[8] However, the conclusion was different for a construct encoding the bee venom protein PLA2, which prevents *Plasmodium* infection through an unknown mechanism.[10] This transgene consistently disappeared in five generations, even after having homogenized the genetic background among competitors. This indicates that the presence of the PLA-2 transgene deteriorates mosquito fitness. Whether this counter-selection was due to PLA2 production or to the chromosomal localization of transgene insertion remains to be clarified.[8]

Overall, these studies confirmed that transgene naturalization is a difficult goal to achieve in *Anopheles* strains even when the effects of genetic drift and of environmental changes are avoided. In these optimal conditions, the major and inescapable source of difficulty stems from the high inbreeding observed within transgenic lines.

Solutions to Naturalization Problems

Sorting Out the Least Costly Transgenes: The Beginning of a Real Solution

Studies on the genetics of adaptation have recurrently shown that mutations generally tend to decrease fitness when expressed in new genetic backgrounds and/or new environments, but that such a fitness cost varies so greatly among mutants than it can occasionally be null.[11-15] Thus, we can reasonably anticipate a variation in fitness cost among transgenes whether this cost arises from the expression of the transgene or from the transgenesis process. As a consequence, common sense recommends increasing the range of mutants to incorporate into transgenesis protocols in order to pick up those associated with the weakest fitness cost. Nonetheless, it is noteworthy that the question of fitness cost intensity per se is far from being the most crucial for the fate of the 'transgenic strategy'. This issue will be discussed in details in the section "Evaluating the Chance of Success of Naturalized Resistance Genes: Formalization and Estimation of the Selective Balance Involved".

Diversifying the Genetic Background of Transgenic Strains: An Achievable Requirement

Although the necessity of using out-bred strains to engineer transgenic mosquitoes has been recognized,[7] genetic drift and low numbers of reproducing adults are so difficult to avoid that an originally out-bred strain is very likely to generate inbred transgenic mosquitoes at release time. A partial solution to alleviate this problem would be to introduce foreign genotypes into transgenic homozygotes just before release time. This may be further optimized if the foreign genotypes are picked up from the vector population targeted by the release strategy. Indeed, such a process will tend to optimize the probability of incorporating the locally adaptive genes into the background of the homozygous transgenic mosquitoes. Replicating such an introgression strategy with parallel population-cages would incorporate alternative wild-type genotypes in the background of the transgenic mosquitoes (Fig. 1B). A final cross among the descendents of these introgressed populations will further diversify the wild-type genotypes of transgenic mosquitoes (Fig. 1B). This crossing protocol looks *a priori* as the most efficient in minimizing the predicted naturalization problems and even reducing the high risk period of naturalization (Fig. 1C). This was indeed one conclusion from an experimental test for the potential of plant population restoration.[6]

Counting on Genetic Drive to Shorten the High Risk Naturalization Period: An Impossible Dream?

The possibility to reduce the critical period of transgene naturalization with the help of genetic drive has received a lot of attention.[1,16-21] A genetic element promoting genetic drive tends to be over-represented in the crossed descendents of carrier and noncarrier parents. Two drive systems have been considered: *Wolbachia*-borne and transposon-borne transgenes. *Wolbachia* are bacteria that parasitize the cells of many Arthropods, have maternal inheritance, and that profoundly affect the reproduction of their host. For instance, *Wolbachia pipientis* affects the

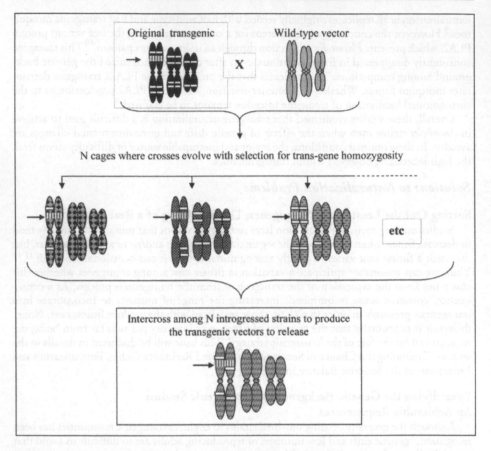

Figure 1B. Using the color conventions of Figure 1A, Figure 1B describes a crossing protocol to apply between the production and release times of *Plasmodium*-resistant mosquitoes. Each population-cage is initially seeded by the offspring of the originally produced transgenic mosquitoes crossed to individuals captured in the population targeted by the future release. Free evolution is allowed in these cages except that only homozygotes for the transgene are allowed to mate. As a consequence, each population will experience a different history and will select for a different genetic background than others. A final cross among resulting strains will thus reincrease the genetic diversity and help remove the remainder deleterious mutations.

fertility of *Culex* and *Aedes* mosquitoes as follows: (i) crosses between *Wolbachia*-infected females and noninfected males are fertile, (ii) crosses between *Wolbachia*-infected males and noninfected females are sterile, and (iii) crosses between two *Wolbachia*-infected parents have variable outcomes depending on the genetic relationships of the bacteria involved. As *Anopheles* mosquitoes are seemingly *Wolbachia*-free in natural populations, the use of a *Wolbachia* from *Culex* or *Aedes* to bear a resistance transgene has been proposed with the idea that the resulting transgenic females will be able to mate with any wild-type male and to transmit the transgene to almost all resulting offspring.[16] However, mathematical analyses showed the occurrence of stringent conditions for *Wolbachia*-induced incompatibilities to actually be able to drive *Wolbachia*-borne transgenes in populations.[17-19] Moreover, if all *Wolbachia* effects remained unchanged in *Anopheles* host, then transgenic males would never be able to produce any viable progeny when mating to wild-females. This poses a serious problem for achieving the necessary reduction in the inbreeding of released transgenic mosquitoes; therefore a reduction in prenaturalization period by such a *Wolbachia*-strategy seems very unlikely.

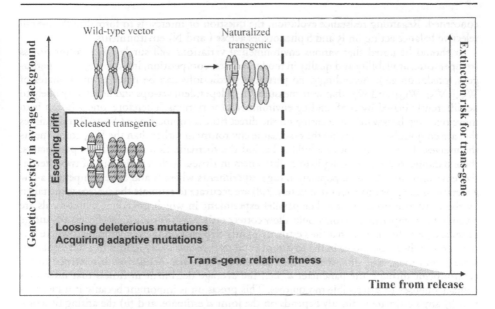

Figure 1C. Using the same conventions as Figure 1A, Figure 1C describes the time-evolution in extinction risks that would be experienced by the introgressed transgenic mosquitoes (obtained from Fig. 1B) once released in the field. Note that the main effect of the crossing protocol would be to reduce the length of the risky period of prenaturalization.

Alternatively, as insect transgenesis generally uses transposon elements,[8,9,22-24] genetic drive via transposon elements was hoped to shorten the naturalization period and to reduce the associated risks of transgene disappearance.[1,20,21] However, a review has recently reduced these hopes to zero by showing that the used transposon elements are at best very poorly remobilized within mosquito genomes.[24]

Evaluating the Chance of Success of Naturalized Resistance Genes: Formalization and Estimation of the Selective Balance Involved

Hereon, we ignore the differences in genetic background between transgenic and native vectors in order to dissect the processes that define the relative fitness of the resistance transgene, the evolution of the resistance phenotype and the epidemiological consequences of this evolution. This section is written with the aim of stimulating discussion among all potential actors involved in a transgenic release project whether they are concerned with laboratory processing, field studies, public health surveys or political decisions. We have thus avoided refining the mathematical models previously developed.[1] Instead, we argue on the biological reality of parameters, their accessibility to estimation in real life, and on their qualitative influence on the conclusions. Therefore, we have tried as much as possible to highlight the bridges that should connect the experimental domain- aimed at identifying and/or modifying the mechanics of *Anopheles-Plasmodium* interactions—to the world in silico—where mathematical models try to capture the evolutionary dynamics of the *'Homo-Anopheles-Plasmodium'* system.

From Biology to Minimal Formalization Able to Forecast Resistance Evolution

The presence and the absence of the risk of *Plasmodium* infection describe two qualitatively distinct environments referred to by the indices I and NI, respectively. The susceptibility and resistance to *Plasmodium* infection define the S and R phenotypes, respectively. Finally, W_{XY} refers to the mean fitness realized by the X phenotype within the Y environment. Given the lack of knowledge on resistance pleiotropy, let us simply assume here that only female fitness is concerned. Regarding resistance evolution, the question of interest is to formalize the overall selective balance acting on R and S phenotypes across I and NI environments.

It should be noted that various environmental variations will surely affect vector fitness (temperature, availability and quality in food resources, competition, infection by other pathogens, predation etc). Accordingly, no pertinent predictions can be made from estimates of W_{SNI}, W_{SI}, W_{RNI} and W_{RI} that were measured in independent set-ups and hence very likely to be differently biased by confounding environmental variations. Therefore, one solution is to move one-step backward, by focusing on the direct fitness comparison of S and R mosquitoes that are conjointly experiencing the exact same environment rather than the W_{XY} components themselves. This is the rationale hidden behind the formalization using c and s parameters: these parameters are measuring R-to-S differences in fitness within either one of two alternative environments. As such, population cage experiments where R and S are competing a few generations long (as those reported in refs. 7,8) are accurate to estimate the average value taken by the resistance fitness cost c. The parallel experiment in which infectious blood-meals are given to the competitive vectors would allow correct estimation of the average fitness advantage of resistance s. Please note that the s estimate tightly depends on the I environment where it is measured: a different choice in the environmental reference is likely to lead to a different estimate. From then, it is noteworthy this I environment is also defined with reference to NI environment through the parameter d; i.e., the average detriment in fitness that infectious contact imposes to susceptible mosquitoes. This precision is important because it underlines that (i) any s estimate indirectly depends on the joint d estimate, and (ii) the arising function $s_{(d)}$ emerges as an inherent property of the environmental conditions chosen as references (i.e., averages in human-blood composition and abiotic parameters, but also in blood-meal concentration in *Plasmodium* gametocytes, parasite genetic composition, and gametocyte infectiousness etc). Therefore, any erroneous appreciation regarding the environmental range experienced by susceptible vectors and the associated variations in vector fitness and/or in parasite fitness would define major sources of errors regarding the estimation and evolutionary importance of the fitness advantage of resistance. Hereafter, the notation $s_{(d)}$ will replace the notation s whenever an evolution in I and NI environmental references is suspected.

For the moment, let us remain with fixed environmental references and describe the formalization using the associated estimates of d, s and c parameters (Fig. 2A). Parameter d measures the fitness detriment that may be imposed by *Plasmodium* infection upon susceptible vectors: $W_{SI} \leq W_{SNI}$ and $W_{SI} = (1-d).W_{SNI}$ with $0 \leq d \leq 1$. Parameter s measures the fitness advantage that resistance confers when a blood-meal is infectious by its infection-blocking effect; thus $W_{RI} = (1 + s).W_{SI}$; with $s \geq 0$. Parameter c measures the fitness cost of resistance when infection risk is absent; thus $W_{RNI} = (1-c).W_{SNI}$ with $0 \leq c \leq 1$. Noting f the probability for a vector to face the risk of *Plasmodium*-infection (i.e., to ingest gametocytes, the parasite stages that are infectious to vectors), the overall selective balance acting on the R/S polymorphism across I and NI environments is simply given by:

$$W_R - W_S = f. [W_{RI} - W_{SI}] + (1-f). [W_{RNI} - W_{SNI}] \tag{1}$$

The pair-wise relationships among female fitness components lead then to rewrite this equation as:

$$W_R - W_S = W_{SNI}. [f.s.(1-d) - (1-f).c] \tag{2}$$

Accordingly, the constraints under which resistance and susceptible phenotypes freely coexist within the local recipient population (i.e., $W_R = W_S$) are defined by:

$$f.s.(1-d) = (1-f).c \tag{3}$$

Alternatively, as transgenic resistant vectors are aimed to be released in viable vector populations, where $W_{SNI} > 0$, the overall fitness difference $[W_R - W_S]$ will have the same sign as the expression $[f.s.(1-d) - (1-f).c]$. Thus, the resistance phenotype will be allowed to locally increase in frequency if and only if $f.s.(1-d) > (1-f).c$.

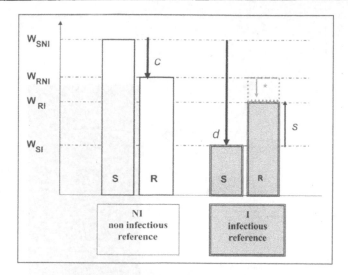

Figure 2A. Fitness effects that are directly induced by the resistance gene: formalization, estimation and evolutionary consequences. Given two environmental references for the vectors, depicting respectively parasite-free (NI) and infectious (I) conditions, this figure pictures a possible outcome for the R-to-S comparison in fitness performance. The fitness detrimental effect of parasite infection in susceptible vectors (pictured by the d-indexed arrow) is an emerging property of this pair of environmental references. From then, fitness comparisons among phenotypes are computed within environments. This allows the most pertinent evaluation of the fitness resistance cost (the arrow indexed by c letter) and fitness resistance advantage (the arrow indexed by s letter), as any confounding environmental variation will simultaneously affect S and R performances. Here, the grey arrow (indexed by a star) illustrates the so-called physiological cost of resistance that is too often confused with that of the resistance fitness cost. Please note that the main effects of this physiological cost are to define a fitness decrease for resistant phenotype in I environment relative to NI, or equivalently to impose an upper-limit to the fitness advantage of resistance. Whenever I and NI environmental references accurately capture the field-environmental variation in selection, a correct evaluation of $\alpha = c/s$ will help forecaste resistance evolution.

Biological Consequences on the Minimal Requirements for Resistance Evolution

The first requirement is obvious: if there is no fitness advantage to resistance ($s = 0$), there must be no fitness resistance cost for the resistance phenotype to be able to persist! In this case, $s = c = 0$: the success of transgenic release strategy depends on the fate of an introduced neutral phenotype, hence on its extinction probability through drift. Both phenotypes would have an average fitness of $W_{RNI} = W_{SNI}$ in parasite-free environments, and of $W_{RI} = W_{SI} = (1-d).W_{SNI} = (1-d).W_{RNI}$ when facing infectious contacts. Interestingly, the value taken by d does not need to be null and estimates two independent quantities at once. In susceptible vectors, it goes on estimating the fitness detriment caused by parasite development, i.e., the effect of virulence of the local *Plasmodium* parasites toward their local vectors. In resistant mosquitoes, because of the emerging relation $W_{RI} = (1-d).W_{RNI}$, the value taken by d quantifies the fitness detriment imposed by the physiological changes allowing them to block parasite development. In other words, here, d measures the physiological cost of a selectively neutral resistance transgene! This highlights once more the absence of synonymy between the physiological and the fitness costs of resistance.[25,26]

Now, it is noteworthy that assuming $d = 0$ does imply $s = 0$ (hence, as above, this implies that resistance evolution relies on the hazardous introduction of a selectively neutral phenotype). Indeed, what can be the fitness advantage of blocking *Plasmodium* development if this infection does not impair fitness? This is a major reason why questions regarding the potential virulence of *Plasmodium* parasite towards their vectors are so crucial.

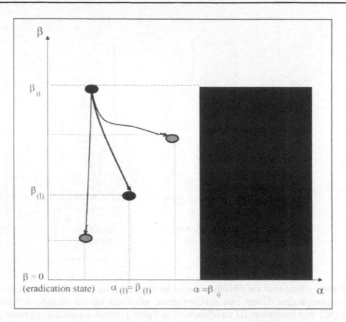

Figure 2B. This figure pictures the conditions of resistance evolution in $\alpha_x \beta$ space. Given the initial values β_0 and α_0, resistance will be allowed to evolve as soon as $\alpha_o < \beta_o$ (i.e., in any point located outside the black-colored area). From then, the expected reduction in human gametocyte burden following the vector evolution of resistance (hence the reduction in *Plasmodium* transmission) will tend to decrease β. Meanwhile, the epidemiological changes induced by vector evolution are very likely to change the average quality of infectious environments (e.g., through changes in human distribution in gametocyte loads and/or genetic composition). In addition, as vector evolution toward parasite resistance will lead to an increase in vectors heterozygous for the transgene, it will change the average response to infection among the contemporaneous phenotypes that are determining the phenotypic references for resistance and susceptibility. As a consequence, the evolution of vector resistance is expected to jointly modify the future values taken by *d* and *s* parameters, and hence by α. In other words, the evolution of resistance in a vector population creates the conditions for a coupled-evolution of α and β. A precise formalization of these α-β coupling-rules will require knowing the functions that determine (i) the dependence of gametocyte load distribution on the probability for human to be infected, (ii) the dependence of the d parameter on human gametocyte load, and (iii) the dominance of s and c parameters in heterozygous vectors. In the absence of this required knowledge, only qualitative conclusions can be made regarding α-β coevolution. In this figure, three qualitative possibilities are pictured with evolution-time following the sense indicated by arrows and evolution starting at the same initial state (α_0; β_0). They are ranked accordingly to the degree of dependence between α and β along their evolutionary course: α and β are rather uncoupled in the example figuring on the left, but evolve in tight interdependence in that figuring on the right. These differences in α-β interdependence did not prevent the occurrence of a time limit l where the whole system reaches an equilibrium defined by $\alpha(l) = \beta(l)$ and $\beta(l) > 0$ (i.e., an absence of parasite eradication), but do nevertheless modify the equilibrium identity through its characteristics proportions in infected humans and vectors.

Finally, let us examine the cases where *Plasmodium* infection does impair the fitness of susceptible vectors ($d > 0$) and resistance does confer a fitness advantage to mosquitoes facing infection ($s > 0$). An extreme case occurs when local humans are all bearing *Plasmodium* gametocytes, such that any blood-meal is infectious for vectors: $f = 1$. In such a case, vector resistance to *Plasmodium* will initially increase in frequency whatever the values taken by *c*, *s* and *d* parameters. Otherwise ($0 < f < 1$), the fate of resistance depends on the comparison between two ratios: $\alpha_{(d)} = c/s_{(d)}$ and $\beta = [f(1-d)] / [1-f]$. An overall selective advantage will favor resistance whenever $\alpha_{(d)} < \beta$, and susceptibility whenever $\alpha_{(d)} > \beta$. One advantage of this formalization is to separate what is specifically related to the fitness comparison among phenotypes within the two chosen environments, from what is inherent to the targeted population. Moreover, this allows the recovery of a classical result in the genetics of adaptation:[27] it is not its fitness cost per se that matters in the evolution of a new phenotype but the suitability between the cost-benefit balance it confers across two given environments of reference ($\alpha_{(d)}$), and the relative importance of those environments both in terms of frequencies (i.e., f terms in β) and of the induced fitness variation in the ancestral phenotype (i.e., d term in β). Finally, this formalization underlines the mechanics of a feed-back linking the evolution of *Plasmodium*-resistance in mosquitoes to the conditions determining its subsequent evolution. Indeed, any increase in *Plasmodium*-resistance in mosquitoes will decrease f- the human-burden of *Plasmodium* gametocytes - and hence necessarily affects the future value taken by β. Complementarily, changes in the quality of the infectious environment are very likely to modify the fitness consequences of *Plasmodium* infection, i.e., to jointly modify the values taken by d, by the relative fitness advantage of resistance $s_{(d)}$, and hence that taken by $\alpha_{(d)}$. In summary, changes in the frequency of resistant vectors will not only modify the transmission patterns from vectors to humans but also the human-environmental referential that counts for the determination of β, d, $s_{(d)}$, $\alpha_{(d)}$ and hence the future issue of α-to β comparisons. This is enough to speculate that- as long as the local parasites do not evolve- the general tendency would be the occurrence of a date-limit l, at which time the dynamics attain equilibrium where α equals β (Fig. 2B).

Returning to real life, the interesting question would be to determine whether or not this equilibrium limit defines a good protection for public health. No pertinent answer can be provided without additional data that explicitly addresses the relationships between the human-distributions in malaria pathologies, in the overall *Plasmodium* burden and in the gametocyte burden, as well as between these human-distributions and the fitness outcomes of parasite-vector interactions (i.e., parasite effects on the fitness of susceptible vectors and the relative benefit of being a resistant vector). In parallel, the overall picture is likely to be modified by parasite evolution. To be favourable for parasite fitness, this evolution should counterbalance the decrease in human-to-vector transmission that has been caused by resistance evolution in vector populations. Among possible evolutionary answers, one solution could be for the parasite to lengthen the time during which successfully infected vectors are infectious for humans, so that a reduction in the density of susceptible vectors would not alter the local parasite density. Another evolutionary solution would be for the parasite to increase its gametocyte burdens in humans, and hence the likelihood for any infected human to participate in the human-to-vector transmission. Other evolutionary responses of the parasite can be proposed. In all cases, the message emerging here is the necessity to fully characterise the interactive relationships that a population of *Plasmodium* parasites jointly entertains with its human-hosts and vectors if we want to be able to evaluate (i) the relative chance for parasite evolutionary answers to vector resistance to be selected for, and (ii) the public health consequences of such potential parasite evolution. To date, required data for such an evaluation are still missing.

Confrontation with an Explicit Model Regarding Resistance Evolution within an Isolated Infinite Population

Boëte and Koella[1] developed an explicit model that was based on the following premises:

1. No parasite evolution;
2. No parasite differences in transmission or in virulence toward mosquito (i.e., each individual parasite that develops within a vector decreases its fitness by a fixed amount);
3. An even distribution in gametocyte burden among the humans infected by Plasmodium parasites at any time t;
4. A frequency of infected humans given by $f(t) = [Ro(t)-1]/[Ro(t) = a/m]$; with a and m representing the mosquito's biting and mortality rates, respectively;
5. Eventuality of partial or total blockage in parasite developments in RR mosquitoes;
6. Poisson distributions in the cumulative parasite burden along mosquito life with similar parasite-dose dependent effects on the fitness detriment of mosquitoes among SS, RS and RR genotypes;

7. A fitness advantage of resistance that arises from an average reduction in parasite burden of RR and RS relatively to SS mosquitoes; so that this advantage also emerges as parasite-dose dependent;

8. RR individuals that either suffer from a fitness resistance cost, or that experience a reduction in their fitness advantage;

9. Equality in dominance levels of the resistance fitness advantage and of the resistance fitness cost;

10. The eventual benefit of genetic drive for resistance transgene.

The conclusions driven by this explicit model are totally congruent with those of our crude and qualitative model. Indeed, in the absence of genetic drive, all parameter combinations result in the occurrence of thresholds in the fitness cost values that allow the evolution of resistance, a diminishing return acting on any evolving resistance, and hence for the very weak probability of *Plasmodium* eradication through modification of the vectorial capacity of a local *Anopheles* population. In addition, this explicit model indicates that only perfect resistance (i.e., so that no single parasite is able to develop within RR mosquitoes) has got a non-null probability of succeeding in a local eradication of *Plasmodium*. From then, the eventual occurrence of genetic drive may facilitate malaria eradication within this isolated population by annihilating the diminishing return effect (i.e., by uncoupling the evolution of α and β).

Deviations Caused by the Co-Ocurrence of Two Vector Species

The very effect of the co-occurrence of two vector species is to tend to uncouple the local evolution of *Plasmodium*-resistance in the vector species of interest from that of human epidemiology (i.e., f), preventing hence β to drop too far from its initial β_0 value. As a result, introduced resistance transgene satisfying $\alpha_0 < \beta_0$ will be more likely to invade the targeted species while parasite-transmission will progressively shift toward the co-occurring vector species. In other words, as soon as two vector species coexist, the population invasion by the resistance transgene is facilitated in the targeted species at the expense of a quasi-null effect on the density of infected humans.

Deviations Caused by Differences in Vector-Parasite Interactions

To this point, we have implicitly assumed that vector resistance and susceptible genotypes respectively display the same interactive issues with all parasite genotypes they encounter. However, this assumption would require direct experimental testing, as nothing ensures that all parasite genotypes interact in the exact same way with all wild-type vector genotypes. This concern is even reinforced by noting that, more often than not, several of the *Plasmodium* spp that infect humans (*P. falciparum*, *P. vivax*, *P. ovale*, and *P. malariae*) coexist and that they can have both positive and negative effects on each other's distribution.[28,29] In addition, experimental data from a rodent malaria model, *P. chabaudi*, and the vector *An. stephensi*, have refuted the assumption under which all parasite genotypes would impose the same detriment in fitness on susceptible vectors.[30] In another experimental infection set-up, one strain of *An. gambiae* selected for complete resistance to *P. cynomolgi* displayed a variable degree in resistance to other related parasite species and strains.[31] Overall, this raises the possibility that *Anopheles-Plasmodium* interaction may obey 'vector genotype x parasite genotype' interactive rules. Therefore, it may be possible that some natural or engineered vector genotypes would better resist infection than others when facing particular parasite genotypes but at the cost to be more easily infected by other parasite genotypes. How would this modify the conclusions of our qualitative model?

As long as the heterogeneity in *Anopheles-Plasmodium* interactive outcomes (i.e., infection success or failure) strictly concerns the wild-type genotypes, it would only induce heterogeneity in the selective advantage conferred by the release-resistance genotypes. Therefore, this would not lead to dramatic changes in our previous conclusions except that the evaluation of the α ratio would be more delicate. Indeed, any infection failure of a wild-type mosquito would nullify the fitness advantage of the released resistance phenotype. Therefore estimation of the average advantage would necessitate an estimate of the proportion of vector-parasite natural encounters that actually result in vector infection. The situation would be more complex if the released-resistance genotype also conferred varying degrees of susceptibility and resistance depending on the *Plasmodium* genotype encountered. In this case, the possibilities for vector resistance evolution (and associated parasite evolution) will depend on the symmetry or asymmetry of the 'vector genotype' x 'parasite genotype' matrix cosigning the infection successes and failures (see refs. 25, 32-34 for details). Therefore, no pertinent conclusion can easily be drawn without previous study on how natural and transgenic *Anopheles* genotypes interact with the *Plasmodium* diversity encountered in the population targeted for a transgenic release.

Nevertheless, let us imagine that the genotype x genotype matrix suggests the possible disappearance of *P. falciparum*, to the benefit of other *Plasmodium* species. Would this lead to a public health benefit? At first glance the answer seems to be 'yes' since *P. falciparum* is presently the most virulent species for humans. Coinfection by other *Plasmodium* species has, however, recurrently been shown to protect humans from severe *P. falciparum* malaria, and it is reciprocally suspected that *P. falciparum* coinfection may provide some protection from serious *P. vivax* pathologies.[28,29] This raises new concerns for the public health outcome following a local change in *Plasmodium* genetic diversity. Whether or not this concern is valid in real life will remain unresolved as long as so little attention is paid to two complementary research axes. The first one concerns the diagnosis of *P. malariae* and *P. ovale* infections: if these infections are rarely reported as defining severe pathologies, is it because of their effectively low virulence or because of a misidentification of the parasite species involved in severe malaria diseases. The second question that would merit more attention concerns the pathology comparisons in mono-versus pluri-specific infections. The increasing prevalence of *P. vivax* and *P. falciparum* coinfection in Thailand could serve as a very good basis for such an investigation.[28,29] Other epidemiological situations can further improve our knowledge regarding the potential impact of *Plasmodium* biodiversity on public health issues. For instance, a replacement of *P. falciparum* by *P. malariae* was suspected to have taken place in some areas of Tanzania, with such replacement being attributed to the efficacy of vector control programs, and potentially the longer duration of human infection by *P. malariae*.[35] Therefore, contemporaneous longitudinal records of malaria pathology (if they exist) may serve as another starting point to directly address the clinical effect of a change in *Plasmodium* genetic diversity.

Deviations Induced When Moving from Population to Metapopulation Levels

In reality, each population is not evolving independently from others. Instead, neighboring populations are connected by migration events that tend to homogenize their genetic composition. This migration effect is counterbalanced by within-population demography promoting genetic drift that tends to make populations genetically diverge from one another, and to impoverish within-population genetic diversity (especially when population size is small). Moreover, population extinction biases this migration/drift balance toward either homogenization or divergence depending on the precise mode of recolonization.[36] The phrase 'metapopulation' takes conjointly into account the effects of migration, genetic drift and extinction/recolonization processes.

Sub-populations in a metapopulation may often differ in the environmental forces pertinent for the phenotype of interest. As vector resistance-to-*Plasmodium* infection is likely to be a selected phenotype, variation in the environmental-induced selection pressures that act upon this phenotype must be considered (i.e., the density of humans that carry *Plasmodium* gametocytes infectious for the vectors, the parasite detrimental effect on the fitness of susceptible vectors, the vector fitness advantage and cost of parasite resistance). As a first approximation, assuming that the resistance fitness advantage and the associated fitness cost are high enough, the evolution of resistance can be forecasted by ignoring drift effects and focusing on migration/selection balance. Several models have investigated this question in the context of resistance to pesticides (see refs. 37-40 and references therein); i.e., a case where among-population-variation in selection pressures is under human control since they correspond to the presence and absence of pesticide treatments. In this context, two key factors affect the possibility of resistance evolution within the metapopulation.[37-40] The first one relates to the geographical distribution of the environmental-induced selection relative to the migration range of the evolving species. This scale is important because it defines the average correlation in the selection experienced by mating individuals and by parents and offspring. The higher these correlations, the easier it is for resistance to evolve in some parts of the metapopulation. The second key factor concerns $\alpha = c/s$ (including the potential but crucial variation in dominance among the cost and advantage that are experienced in heterozygotes, see ref. 40 for details). Assuming an absence of parasite evolution in response to that of the *Plasmodium*-resistance in vectors, these conclusions can be extended to the case of mosquito-resistance to *Plasmodium*. Here, the pertinent environmental variation concerns the risk of *Plasmodium*-infection that the vector species of interest faces. Therefore, even if this risk is not under human control, any factor that will affect the correlation of such a risk among mating individuals and/or among parents and offspring will jeopardize the evolution of *Plasmodium*-resistance within a vector metapopulation. Quantitative forecasting will require not only knowledge of the geographical distribution of *Plasmodium* gametocytes, but also that of *Plasmodium* infectiousness and virulence toward wild-type vectors (i.e., components affecting the intensity of resistance fitness advantage).

Recap of the Biological Data Required to Correctly Forecast Vector and Disease Evolution

Recall that no resistance evolution can ever occur following the release of laboratory mosquitoes if *Plasmodium* infection does not impair the fitness of susceptible vectors. Direct tests of this assumption that include estimates of parasite detrimental effects on vector fitness remain nevertheless rare, with the exception of the work of Hurd and her collaborators.[41-43] These authors reported drastic reductions (from 15% to 48%) in mosquito fecundity in laboratory-controlled infections of *An. stephensi* and *An. gambiae* by *P. yoelii nigeriensis*, and in Tanzanian natural infections of *An. gambiae* by *P. falciparum*. This detrimental effect on female fecundity was found to last over at least three consecutive gonotrophic cycles, and to affect both strongly and weakly parasitized females (harboring >75 oocysts or ≈ 4 oocysts). Moreover, the extent of the reduction in fecundity was similar in strongly and weakly parasitized females during the 2nd and 3rd gonotrophic cycles (≈ 20% to 25% reduction). This suggests that there is weak parasite-dose dependence in the fitness detriment imposed by *Plasmodium* infection on their vectors. This would in turn suggest that there is a weak β-diminishing return on both α and the long-term evolution of *Plasmodium*-resistance in vector populations, and hence a high potential for a transgenic release to actually reduce *Plasmodium* burden in humans.

It nevertheless remains pivotal to test whether such 'favorable' characteristics tend to be the general rule among all the local possibilities of *Plasmodium - Anopheles* genetic combinations. As far as we know, this issue has not yet received attention. However, its importance appears reinforced by the recurrent reports of highly aggregated distribution of parasites among naturally infected vectors.[44] Such aggregation can either result from a high heterogeneity in the infectiousness among local blood-meals, or from a local variation in vector susceptibility to *Plasmodium* infections. Interestingly both possibilities are indirectly supported by our current knowledge of malaria biology. On the one hand, it has been proposed that, in some localities, 20% of a local human infectious reservoir may be responsible for 80% of malaria transmission.[45] On the other hand, naturally occurring *Anopheles* resistance to *Plasmodium* infection has begun to be identified.[47-49] Both possibilities would have dramatic consequences on the α-to-β comparisons that regulate the long-term evolution of the released transgenic resistance within vector populations. Indeed, they will respectively reduce β, through a reduction in the *f* probability to face infection risk, and increase α by reducing the fitness advantage of the released-resistance since natural vectors may be as able to resist infection as the released ones. Therefore, no pertinent forecast of the local evolution of a released *Plasmodium*-resistance phenotype will ever be possible without previously knowing how natural *Anopheles* genotypes actually interact with *Plasmodium* ones in the locality chosen for this future release.

The need for data that document vector-parasite interactions appears even more crucial when addressing the probability for a vector to take an infectious blood-meal (parameter *f,* see section "From Biology to Minimal Formalization Able to Forecast Resistance Evolution"). Gametocyte burdens greatly vary with region of study and with human history of exposure and tend to reflect the overall asexual parasite density:[50,51] infections in the younger and therefore less immune children tend to produce higher densities of gametocytes than adults.[52-54] However, although increasing gametocyte density tends to result in greater infectiousness to mosquitoes, high gametocyte densities do not guarantee high mosquito infection rates.[55-58] Indeed, cryptic infected humans, apparently bearing no or very few apparent gametocytes, are capable of infecting mosquitoes and may contribute to a very significant proportion of the human transmission reservoir.[57-59] Such variability in human-to-mosquito transmission suggest several sources of variation that must be identified and taken into account in explicit evaluation of models of the transgenic strategy. So far, two sources of variation have been identified and a third may be hypothesized.

- First, the periodicity and intensity of malaria transmission are known to conjointly affect the distribution frequencies in gametocyte carriage among human age classes. In endemic low transmission areas, parasite prevalence rates are similar across all age groups and gametocytes are similarly frequent in all age groups.[60] As the transmission intensity increases, both overall and gametocyte parasite prevalence rates decrease with human age, reflecting the acquisition of immunity.[52,53] By contrast, in regions where malaria is epidemic, gametocytes are found at high frequency in all age groups and remain largely absent during inter-epidemic periods. Interestingly, these characteristics are entirely congruent with the assumption that *Plasmodium* parasite may be able to modify its human-borne life-stages to adapt to variations in vector availability.
- Second, the duration of gametocyte positivity in an individual infection varies with age and disease severity.[61] Both these factors are likely to vary not only with transmission intensity, but also with the local parasite specific diversity, and importantly, the efficacy of local health networks. Once more, this suggests a non-null capacity of *Plasmodium* parasites for finding adaptive answers to variations in transmission constraints.
- Third, we can speculate on the occurrence of geographical differences in the match-pattern of *Plasmodium* and *Anopheles* genotypes that allow vector infection. Interestingly, two experimental studies provide indirect support for this.[30,31]

Thus, with a minimum of knowledge on the local transmission intensity, crude estimates of *f* can be made. How these characteristics will be altered by a decrease in transmission is uncertain, but such data should be available from previous studies where transmission has been reduced by extrinsic methods (e.g., bednet studies). Determining local parasite-vector adaptation and/or local adaptive capacity of parasites to respond to epidemiological changes may prove more problematic but potentially more important.

The Complex Biological Reality of Malaria: How Does the Acquired Knowledge Relate to the Pivotal Parameters of Resistance Evolution?

The Specific Diversity in Malaria Vectors

The engineering of resistant transgenic mosquitoes tends to focus on *Anopheles gambiae,* one of the major malaria vectors in the Afro-tropical region where the virulent *P. falciparum* is mostly found. However, more than one hundred *Anopheles* species have been described as malaria vectors worldwide,[62] and, even in the Afro-tropical region, the local co-occurrence of diverse vector species of *P. falciparum* looks to be the rule (Fig. 3). Furthermore, the so-called

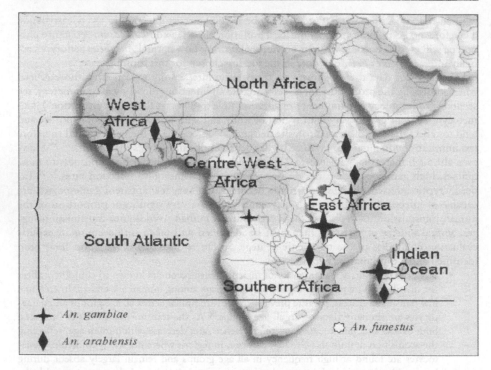

Figure 3. Biological diversity involved in malaria transmission: a neglected area of critical importance. More than one hundred *Anopheles* species have been described worldwide as vectors for malaria parasites.[62] Nevertheless, refractory transgenic mosquitoes mainly concern *An. gambiae* that is considered as the "main vector" in Afro-tropical region where *P. falciparum* is responsible for most malaria scourge. But: i) *An. gambiae* constitutes a species complex including *An. gambiae s.s.*, *An. arabiensis*, *An. quadriannulatus*, *An. melas* and *An. merus.*[69] ii) Other main vectors of *P. falciparum* frequently coexist with *An. gambiae s.s.* in this area. (See the figure below for a rough description restricted to "main" vectors). iii) Other *Plasmodium* species (*P. vivax*, *P. malariae*, *P. ovale*) can also coexist with *P. falciparum* in Afro-tropical sites.[70] In the face of these observations, the major issue of transgenic release looks to focus on two interrelated questions: What could be the consequences of a release of refractory *An. gambiae s.s.* strains in Afro-tropical zone on the dynamics of the evolutionary relationships that *P. falciparum* entertains there with (i) its other vectors, and with (ii) the other malaria parasites?

An. gambiae is a species complex regrouping gene-pools that evolve in relative independence from one another (Fig. 3). This raises different concerns regarding the potential of the transgenic release strategy. Firstly, there is no a priori guarantee concerning the biological compatibility between the engineered-strains to be released in a locality and the *P. falciparum* vectors living there. Secondly, the likely co-occurrence of competing vector species leads one to expect that the transgenic release strategy would at best modify the parasite-transmission-pattern with an uncertain impact on public health (see section "Deviations Caused by the Co-Ocurrence of Two Vector Species"). Circumventing these two difficulties by targeting several major vector species simultaneously would make the preparatory and production phases increasingly onerous, even when focusing on restricted geographical areas.

Such difficulty is exemplified by the joint investigation of malaria transmission and *Anopheles* diversity in the Senegalese villages of Dielmo and Ndiop, located 5km apart.[63,64] In Dielmo, the year-round presence of *An. funestus* guarantees continuity in malaria transmission. Six other *Anopheles* species were also identified, among which three were involved in malaria transmission to differing extents according to the season. Even more interestingly, *Anopheles* species that did not appear to be involved in malaria transmission in Dielmo were classified as good malaria vectors elsewhere. As a result, the human-to-vector transmission pattern does not only involve distinct vector species within localities but these patterns are also heterogeneous along seasons and over African localities. This is not an exceptional situation as a comparable scenario was observed in Papua New Guinea, where anopheline diversity involves *An. koliensis*, three species of *An. punctulatus* complex, and six of *An. farauti* complex.[65] All these species are malaria vectors with both seasonal and geographical variation in their relative importance due to their variable degree of zoophily and ecology.[65,66]

The involvement of a vector species in malaria transmission actually defines the selection pressures acting on *Plasmodium*-resistance evolution. As a result, seasonal and geographical changes are likely to reduce the average correlation in the selection experienced among subsequent generations and among mating individuals that have previously migrated. This is worthy of note since such a reduction in correlation jeopardizes the evolution of resistance at the metapopulation scale (see section "Deviations Induced When Moving from Population to Metapopulation Levels"). This may be one of the reasons explaining why naturally arisen resistance to *Plasmodium* infections seems so rare in vector populations.

Wild Anopheles Vectors Have Naturally Developed Resistance to Plasmodium Infection

It has been argued that the lack of *Plasmodium* melanization by natural vectors to *Plasmodium* infection[67] indicates that the resistance fitness cost overcomes the resistance fitness advantage in nature.[68] However, melanization is not the only immune response of Insects, and alternative defense mechanisms have been shown to protect *Anopheles* natural genotypes from *Plasmodium* infections.[46-49] Given the lack of biological data in virtually all domains necessary for quantifying the fate of transgenic release (e.g., intensities of resistance fitness advantage and cost, distribution probability for the vector risk to encounter a virulent parasite, variation in the intensity of parasite virulence toward vectors, possibility that resistance to some parasite genotypes facilitates the transmission of others etc.), it would be of great help to start dissecting the evolutionary factors that actually regulate the evolution of such naturally occurring resistance mechanisms. Such investigations would certainly allow the incorporation of more realistic parameters in the explicit models aimed at evaluating the fate and epidemiological consequences of the release of laboratory-engineered resistant mosquitoes.

Conclusion on the Most Likely Results of the Release of Laboratory Resistant Anopheles

First of all, it is noteworthy that the majority of the present discussion does not strictly concern the particular case where mosquito resistance arises from transgenesis. The only exception is the discussion on the possible use of genetic drive to reduce the high risk period of naturalization (see section "Transgene Naturalization in Laboratory Experiments: The Actual Dimension of Inbreeding"). Otherwise, the entire discussion would be exactly the same whether the subject was transgenic resistant mosquitoes or selected pools of mosquitoes bearing naturally occurring resistant genes.

From this point, investigating the various sources able to affect the evolutionary fate of a released transgenic vector resistant to *Plasmodium* has led to a few qualitative conclusions. First, the major risks for transgene extinction are indeed expected to occur very soon after release (Fig. 1). Second, these major risks are more due to the laboratory-origin of the released mosquitoes than to the resistance characteristic it confers (Fig. 1). Therefore, even if some of these risks may be prevented before transgenic release (see section "Transgene Naturalization in Laboratory Experiments: The Actual Dimension of Inbreeding"), the more likely outcome of a transgenic release strategy would be a rapid disappearance of the released transgene, inducing thus no visible change in malaria epidemiology, and hence neither public health improvement nor any medical and/or environmental risks. Third, possible consequences were pinpointed in the unlikely case where a released resistant genotype succeeds in persisting long enough to actually modify the vector competence of the targeted species. Interestingly, the major expected outcome will rely on the specific diversity of the vectors, the parasites, or both (see sections "Deviations Caused by the Co-Ocurrence of Two Vector Species", "Deviations Caused by Differences in Vector-Parasite Interactions" and "Deviations Induced When Moving from Population to Metapopulation Levels"). Indeed, the likely outcomes of a persistent resistance transgene within a vector population would be a passive shift in the vector locally used by *Plasmodium* parasites and/or a modification in the *Plasmodium* genetic diversity that is locally circulating, either within or among *Plasmodium* species. Evaluating whether or not these outcomes would modify public health criteria would require data acquisition regarding eventual genetic variability of *Plasmodium* parasites in virulence-transmission tradeoffs.

Overall, it seems that the major predictable outcome of the transgenic release strategy would not concern public health but fundamental science! Indeed, a correct evaluation of the probability and consequences of the success of such a strategy will require reconsidering *Plasmodium* biology and genetics to its very roots, given the recurrent calls for the acquisition of remaining unknown data throughout the present discussion. For 50 years, malaria epidemiology has been based on the Macdonald's quantitative epidemiological model that ignores genetic variability among malaria parasites and their vectors. Since most of immediate consequences of transgenic release depend on parasite and vector genetic diversity, this classical model is no longer sufficient. On an optimistic note, as transgenic release strategy is so well suited for the media, we wager that the acquisition of the required but still missing data will soon benefit from the increasing attention. If this bet is correct, then the information pertinent to understanding the mechanics determining the dynamics of the malaria scourge will unquestionably increase.

References

1. Boëte C, Koella JC. A theoretical approach to predicting the success of genetic manipulation of malaria mosquitoes in malaria control. Malar J 2002; 1(1):3.
2. Benedict MQ, Robinson AS. The first releases of transgenic mosquitoes: An argument for the sterile insect technique. Trends Parasitol 2003; 19(8):349-355.
3. Tripet F, Touré YT, Dolo G et al. Frequency of multiple inseminations in field-collected Anopheles gambiae females revealed by DNA analysis of transferred sperm. Am J Trop Med Hyg 2003; 68(1):1-5.
4. Davidson G, Odetoyoinbo JA, Colussa B et al. A field attempt to assess the mating competitiveness of sterile males produced by crossing two member species of the Anopheles gambiae complex. Bulletin WHO 1970; 42:55-67.
5. Hufford KM, Mazer SJ. Plant ecotypes: Genetic differentiation in the age of ecological restoration. Trends Ecol Evol 2003; 18(3):147-155.
6. Veerger P, Sondered E, Oubordg N. Introduction strategies put to the test: Local adaptation versus heterosis. Conserv Biol 2004; 18:812-821.
7. Catteruccia F, Godfray CJ, Crisanti A. Impact of genetic manipulation on the fitness of Anopheles stephensi mosquitoes. Science 2003; 299:1225-1227.
8. Moreira LA, Wang J, Collins FH et al. Fitness of Anopheline mosquitoes expressing transgenes that inhibit Plasmodium development. Genetics 2004; 166:1337-1341.
9. Ghosh AK, Ribolla PEM, Jacobs-Lorena M. Targeting Plasmodium ligands on mosquito salivary glands and midgut with a phage display peptide library. Proc Natl Acad Sci USA 2001; 98:13278-13281.
10. Moreira LA, Ito J, Ghosh AK et al. Bee venom phospholipase inhibits malaria parasite development in transgenic mosquitoes. J Biol Chem 2002; 277(43):40839-40843.
11. Lenski RE. Experimental studies of pleiotropy and epistasis in Escherichia coli. I. Variation in competitive fitness among mutants resistant to virus T4. Evolution 1988; 42:425-432.
12. Bouma JE, Lenski RE. Evolution of a bacteria/plasmid association. Nature 1988; 335:351-352.
13. Cohan FM, King EC, Zadawski P. Amelioration of the deleterious pleiotropic effects of an adaptive mutation in Bacillus subtilis. Evolution 1994; 48:81-95.
14. Lenski RE, Rose MR, Simpson SC et al. Long-term experimental evolution in Escherichia coli. I. Adaptation and divergence during 2,000 generations. Am Nat 1991; 138:1315-1341.

15. Guillemaud T, Lenormand T, Bourguet D et al. Evolution of resistance in Culex pipiens: Allele replacement and changing environment. Evolution 1998; 52:443-453.
16. Riehle MA, Srinivasan P, Moreira C et al. Towards genetic manipulation of wild mosquito populations to combat malaria: Advances and challenges. J Exp Biol 2003; 206:3809-3816.
17. Curtis CF, Sinkis SP. Wolbachia as possible means of driving genes into populations. Parasitology 1998; 116:S111-S115.
18. Turelli M, Hoffmann AA. Microbe-induced cytoplasmic incompatibility as a mechanism for introducting transgenes into arthropod populations. Insect Mol Biol 1999; 8:243-255.
19. Rasgon JL, Scott TW. Impact of population age structure on Wolbachia transgene driver efficacy: Ecologically complex factors and release of genetically modified mosquitoes. Insect Biochem Mol Biol 2004; 34:707-713.
20. Kiszewski AE, Spielman A. Spatially explicit model of transposon-based genetic drive mechanisms for displacing fluctuating populations of anopheline vector mosquitoes. J Med Entomol 1998; 35:584-590.
21. Ribeiro JMC, Kidwell MG. Transposable elements as population drive mechanisms: Specification of critical parameter values. J Med Entomol 1994; 31:10-15.
22. Ito J, Ghosh LA, Moreira LA et al. Transgenic anopheline mosquitoes impaired in transmission of a malaria parasite. Nature 2002; 417:452-454.
23. Moreira LA, Jacobs-Lorena M. Transgenic mosquitoes for malaria control: Progresses and challenges. Neotrop Entomol 2003; 32:531-536.
24. O'Brochta DA, Sethuraman N, Wilson R et al. Gene vector and transposable element behavior in mosquitoes. J Exp Biol 2003; 206:3823-3834.
25. Coustau C, Chevillon C, ffrench-Constant R. Resistance to xenobiotics and parasites: Can we count the cost? Trends Ecol Evol 2000; 15:378-383.
26. Coustau C, Théron A. Resistant or resisting: Seeking consensus terminology. Trends Parasitol 2004; 20:209-210.
27. Nagylaki T. Conditions for the existence of clines. Genetics 1975; 80:595-615.
28. Snounou G, White NJ. The coexistence of Plasmodium: Sidelights from falciparum and vivax malaria in Thailand. Trends Parasitol 2004; 20:333-339.
29. Zimmerman PA, Melhotra RK, Kasehagen LJ et al. Why we do need to know more about Plasmodium species infections in humans? Trends Parasitol 2004; 20:440-447.
30. Ferguson HM, Read AF. Genetic and environmental determinants of malaria parasite virulence in mosquitoes. Proc R Soc Lond B Biol Sci 2002; 269(1497):1217-1224.
31. Collins FH, Sakai PK, Vernick PD et al. Genetic selection of a Plasmodium-refractory strain of the malaria vector Anopheles gambiae. Science 1986; 234:607-610.
32. Frank SA. Statistical properties of polymorphism in host-parasite genetics. Evolutionary Ecology 1996; 10:307-317.
33. Frank SA. Problems inferring the specificity of plant-pathogen genetics. Evolutionary Ecology 1996; 10:323-325.
34. Parker M. The nature of plant-parasite specificity. Evolutionary Ecology 1996; 10:319-322.
35. Bruce-Chwatt LJ. A longitudinal survey of natural malaria infection in a group of West African adults. West Afr Med J 1963; 12:141-173.
36. Wade MJ, McCauley DE. Extinction and recolonisation: Their effects on the genetic differentiation of local populations. Evolution 1990; 42:995-1005.
37. Mani GS. Evolution of resistance with sequential application of insecticides in time and space. Proc R Soc Lond B Biol Sci 1989; 238:245-276.
38. Lenormand T, Raymond M. Resistance management: The stable zone strategy. Proc R Soc Lond B Biol Sci 1998; 265:1985-1990.
39. Vacher C, Bourguet D, Rousset F et al. Modelling the spatial configuration of refuges for sustainable pest control/ a case study of Bt-cotton. J Evol Biol 2003; 16:378-387.
40. Vacher C, Bourguet D, Rousset F et al. High dose refuges strategies and genetically modified crops - Reply to Tabashnik et al. J Evol Biol 2004; 17:913-918.
41. Hurd H. Manipulation of medically important insect vectors by their parasites. Annu Rev Entomol 2003; 48:141-161.
42. Hurd H, Hogg JC, Renshaw M. Interactions between bloodfeeding, fecundity and infection in mosquitoes. Parasitol Today 1995; 11:411-416.
43. Ahmed AM, Maingon R, Romans P et al. Effects of malaria infection on vitellogeneisis in Anopheles gambiae during two gonotrophic cycles. Insect Mol Biol 2001; 10:347-356.
44. Pringle G. A quantitative study of naturally-acquired malaria infected in Anopheles gambiae and Anopheles funestus in a highly malarious area of East Africa. Trans R Soc Trop Med Hyg 1966; 60:626-632.

45. Woolhouse MEJ, Dye C, Etard J-F et al. Heterogeneities in the transmission of infectious agents: Implications for the design of control programs. Proc Natl Acad Sci USA 1997; 94:338-342.
46. Blandin S, Shia S-H, Moita LF et al. Complement-like protein TEP1 is a determinant of vectorial capacity in the malaria vector Anopheles gambiae. Cell 2004; 116:661-670.
47. Christophides GK, Vlachou D, Kafatos FC. Comparative and functional genomics of the innate immune system in the malaria vector Anopheles gambiae. Immunol Rev 2004; 198:127-148.
48. Dimopoulos G, Christophides GK, Meisler S et al. Genome expression analysis of Anopheles gambiae: Responses to injury, bacterial challenge, and malaria infection. Proc Natl Acad Sci USA 2002; 99(13):8814-8819.
49. Osta MA, Christophides GK, Kafatos FC. Effects of mosquito genes on Plasmodium development. Science 2004; 303:2030-2032.
50. Thomson D. A research into the production, life and death of crescents in malignant tertian malaria, in treated and untreated cases, by an enumerative method. Ann Trop Med Parasitol 1911; 5:57-82.
51. Christophers R. The mechanics of immunity against malaria in communities living under hyper-endemic conditions. Indian J Med Res 1924; 12:273-294.
52. Wilson DB. Rural hyper-endemic malaria in the Tanganyika territory. Trans R Soc Trop Med Hyg 1936; 29:583-618.
53. Davidson G, Draper CC. Field studies of some of the basic factors concerned in the transmission of malaria. Trans R Soc Trop Med Hyg 1953; 47:522-535.
54. Davidson G. Further studies of the basic factors concerned in the transmission of malaria. Trans R Soc Trop Med Hyg 1955; 49:339-350.
55. Jeffery GM, Eyles DE. Infectivity to mosquitoes of Plasmodium falciparum as related to gametocyte density and duration of infection. Am J Trop Med Hyg 1955; 4:781-789.
56. Muirhead-Thomson RC. Low gametocyte-threshold of infection of Anopheles with Plasmodium falciparum. Br Med J 1954; 1:68-70.
57. Muirhead-Thomson RC. The malaria infectivity of an African village population to mosquitoes (Anopheles gambiae). Am J Trop Med Hyg 1957; 6:971-979.
58. Boudin C, Olivier M, Chiron JP et al. High human malarial infectivity to laboratory-bred Anopheles gambiae in a village in Burkina Faso. Am J Trop Med Hyg 1993; 48:700-706.
59. Bonnet S, Gouagna CL, Paul RE et al. Estimation of malaria transmission from humans to mosquitoes in two neighbouring villages in south Cameroon: Evaluation and comparison of several indices. Trans R Soc Trop Med Hyg 2003; 97:53-59.
60. Earle WC, Perez M, del Rio J et al. Observations on the course of naturally acquired malaria in Puerto Rico. PR J Public Health Trop Med 1939; 14:391-406.
61. Bousema JT, Gouagna LC, Drakeley CJ et al. Plasmodium falciparum gametocyte carriage in asymptomatic children in western Kenya. Malar J 2004; 3:3.
62. Service MW. Mosquitoes (Culicidae). In: Lane RP, Grosskey RW, eds. Medical Insects and Arachnids. London: Chapman & Hall, 1993:120-240.
63. Fontenille D, Lochouarn L, Diagne N et al. High annual and seasonal variations in malaria transmission by anopheline and vector species in Dielmo, a holoendemic area in Senegal. Am J Trop Med Hyg 1997; 56:247-253.
64. Fontenille D, Lochouarn L, Diatta M et al. Four years' entomological study of the transmission of seasonal malaria in Senegal and the bionomics of Anopheles gambiae and A. arabiensis. Trans R Soc Trop Med Hyg 1997; 91:647-652.
65. Beebe NW, Cooper RD. Distribution and evolution of the Anopheles punctulatus group (Diptera: Culicidae) in Australia and Papua New Guinea. Int J Parasitol 2002; 32:563-574.
66. Burkot TR, Graves PM, Paru R et al. Mixed blood feeding by the malaria vectors in the Anopheles punctulatus complex (Diptera: Culicidae). J Med Entomol 1988; 25:204-213.
67. Schwartz A, Koella JC. Melanization of Plasmodium falciparum and C-25 Sephadex beads by field-caught Anopheles gambiae (Diptera: Culicidae) from Southern Tanzania. J Med Entomol 2002; 39(1):84-89.
68. Boëte C, Koella JC. Evolutionary ideas about genetically manipulated mosquitoes and malaria control. Trends Parasitol 2003; 19(1):32-38.
69. Coetzee M, Craig M, le Sueur D. Distribution of African mosquitoes belonging to the Anopheles gambiae complex. Parasitol Today 2000; 16:74-77.
70. Garnham PCC. Malaria parasites and other haemosporidia. Oxford: Blackwell Scientific Publications, 1966.

CHAPTER 11

The Genetics of Vector-Host Interactions:
Alternative Strategies for Genetic Engineering for Malaria Control

Willem Takken* and Carlo Costantini

As long as behavioural genetics remains a scientific backwater, much of the genome sequence will look like uninterruptible gibberish.
—*Of Flies and Men*, by Dean H. Hamer, *Scientific American*, June 1999

Abstract

Malaria transmission is accomplished by the innate behavioural trait of mosquitoes to ingest vertebrate blood required for egg production. As human malaria parasites are, by definition, circulating between humans and certain anopheline species, disruption of mosquito-human contact will effectively inhibit transmission of the malaria parasite. Here we explore factors that affect mosquito-host interactions to assess how this process can be exploited to reduce malaria transmission. Host preference in mosquitoes is genetically controlled, and it is argued that a change in host preference could result in less human biting and in reduced parasite transmission. The effect of this is being demonstrated using the vectorial capacity equation, in which the human biting index and mosquito survival are represented. It is argued that effective malaria control strategies should be based on a reduction of human biting preference coupled with reduced survival. Strategic interventions based on behavioural manipulation and ecological change may affect the biting fraction of the vector population to such an extent that the vectorial capacity is significantly affected. In some cases this may require genetic modification of organisms (GMO) technology, but mechanical or physical techniques should also be considered.

Introduction

The human *Plasmodia*, the causative agents of one of the deadliest diseases on Earth: malaria, are unquestionably among the most successful of the vector-borne parasites, overcoming the natural resistance mechanisms of their vertebrate and arthropod hosts, and showing strong resilience against conventional methods of disease control. This results in more than one million deaths every year due to this disease. Genetic variability of the *Plasmodia* and their association with a relatively small group of mosquitoes provide the key to interpret this success. As sexual recombination in the *Plasmodia*, and hence the mechanism insuring a higher degree of genetic variability, is accomplished in the mosquito midgut, mosquito fitness and behaviour are of critical importance for the parasite. The concept of genetic engineering technologies as an alternative method for malaria control is dominated by the notion of manipulation of vector competence through modification of the mosquito natural immunity against the parasite.[1,2]

*Corresponding Author: Willem Takken—Laboratory of Entomology, Wageningen University, P.O. Box 8031, 6700 EH Wageningen, The Netherlands. Email: willem.takken@wur.nl

Genetically Modified Mosquitoes for Malaria Control, edited by Christophe Boëte.

Here we argue that other traits of anopheline mosquitoes could represent adequate targets for intervention by genetic manipulation, and might result as effective means for the interruption of malaria parasite transmission. This argument is encouraged by the recent publication of the malaria mosquito complete genome sequence and the continuing development of high-throughput genomic technologies, which everyone hopes will provide in the future the technological basis to investigate and identify novel targets for intervention.[3] However, as a judicious reminder evoked in the citation opening this chapter, we need to take into account and relate such technological advances with the grassroots biology of the vectors, hence to a deeper understanding of their field ecology and behaviour, or our efforts are bound to fail.[4] As many studies on insect transgenesis originate in the laboratory, it is obvious that the transfer of this technology from the bench to the field requires specific attention lest one ends up with a mosquito that has lost several of its natural traits.[5] It is didactic and perhaps farsighted that such a reminder comes from the community of scholars studying the fruit fly *Drosophila melanogaster* as a model organism (whose genome sequence was completed well before that of the malaria mosquito), as their intimate knowledge of the biology and genetics of this species is arguably unparalleled in the animal kingdom.[6,7] Intrinsic and extrinsic factors determine mosquito fitness and vector competence.[8] Behavioural traits like host preference, diurnal rhythms and locomotion affect the uptake and spread of the parasite. The implications of these aspects of vector biology with respect to malaria transmission and control are now discussed in the light of the proposed GMO technology for malaria control. Different aspects of mosquito behaviour are presented, followed by a discussion about whether genetic modification of behavioural traits might be considered as a potential strategy for disease control radically different from the strategies based on vector competence.

Vector Olympics

The success of a malaria parasite can be measured by the rate of its spread through a human community, expressed as Basic Reproductive Rate.[9] In practical terms, we measure this through the vectorial capacity,[10] a derivative of the basic reproductive rate (see Box 1). Apart from a demographic factor expressing the mean longevity of the vector population (p), this equation also contains a behavioural component (a) which is the frequency of mosquito bites on humans, which in turn depends on the proportion of the vector population selecting humans as a blood host. This factor is squared, because the mosquito needs to bite two subsequent times to transmit the parasite, first to become infected, and second to pass the parasite on to another human host, after having allowed for the completion of the parasite sporogonic cycle in the mosquito. This extrinsic incubation time (n) is also dependent on the behaviour of the mosquito: should she choose to spend a lot of time in environments having favourable micro-climatic conditions (constant and relatively high temperatures), the development of the parasite from ookinete to sporozoites will occur faster compared to siblings remaining at lower ambient temperatures.

Another factor to be considered is the natural susceptibility of the mosquito for parasite development, expressed as vector competence. This is determined by the genetic make up of both the vector and the parasite, and possibly explains why only some 60 anopheline species are suitable vectors for human *Plasmodia*.[11,12] A "good" malaria vector is therefore characterized by high longevity, a high degree of anthropophily, and a tendency to seek shelter in an environment with relatively high ambient temperatures while digesting the blood meal, as well as a high susceptibility for parasite development. Less successful vector species fail to have some of these characters. Yet, some anopheline vectors expressing a favourable combination of these parameters, such as species of the *An. dirus* complex in Southeast Asia, are not among the world champions of malaria transmission, if judged by the incidence of infections they cause, because they live in forested areas at the margins of the human environment, hence are not favourably impacted by human modifications of the natural habitat.[13] The degree of sinanthropomorphism, or anthropophily in its loosest meaning, is therefore another important biological trait insuring the success of an anopheline species as a malaria vector. This is not accounted for in the

Box 1. The vectorial capacity equation

Vector capacity is the daily rate at which new human infections arise due to the introduction in a malaria-free area of a single gametocyte carrier, i.e., the malaria multiplying potential in the human population due to the vector.

$$c = \frac{ma^2p^n}{-\ln p}$$

m - total number of *Anopheles* per person
a - frequency of bites on humans per vector per day
p - vector mean daily survival rate
n - *Plasmodium* extrinsic incubation duration (in days)

formulation of vectorial capacity, but it is arguably one of the reasons why the vectors in sub-saharan Africa champion malaria transmission and have the regretful repute of accounting for 90% of the world malaria burden.

It follows that strategies for malaria control should be directed to impact some or all of these factors. Usually this is accomplished by spraying of insecticides that cause reduced longevity or by the use of insecticide-impregnated bed nets that reduce indoor biting and resting behaviour. These interventions can cause a reduction in parasite transmission, but have not been shown to affect the genetically-determined traits of anthropophily and vector competence.

Feeding Behaviour

All anautogenous mosquitoes require vertebrate blood for egg production. Some species are opportunistic in this behaviour, and feed on any type of blood host, provided sufficient quantities of blood can be ingested to permit egg development.[14] Others, by contrast, have evolved oligotrophic habits and feed on a limited number of hosts or even on a single host species such as *Deinocerites dyari*.[15] Most malaria vectors belong to the first category, but several important vectors feed preferentially on humans. To this group of anopheline mosquitoes belong *Anopheles gambiae sensu stricto*, *An. funestus* and *An. nili* in Africa and *An. fluviatilis* species S in Asia.[16-20] These species have evolved a strong association with humans by adapting to human habitation as feeding and resting ground, finding shelter inside people's dwellings and biting preferentially at times when the host is asleep.[21,22] The acquired endophilic and endophagic feeding behaviours accidentally enhance the mosquito's survival because the human home offers a relatively stable environment with protection from predators and extreme meteorological events. Furthermore, for endophilic and anthropophilic mosquito species such as *An. gambiae s.s.* the human host is always close by, unlike outdoors, where host availability can be haphazard causing the insect to loose precious energy during host searching, thereby augmenting the general fitness of such species.[22,23] Nevertheless, in specific circumstances, normally-endophagic mosquitoes can bite excessively outdoors, presumably in response to ambient conditions.[24]

The degree of anthropophily, i.e., the intrinsic or endogenous preference for feeding on human hosts, is an important character in the equation of malaria transmission. This character has a genetic basis, as demonstrated by experiments selecting for higher or lower degrees of anthropophily than baseline strains in species of the *An. gambiae* complex (H.V Jamet (Pates), PhD dissertation, London 2002). The evolution of anthropophily might have followed different paths in separate species, and at least three processes can be suggested: (i) shift from primitive simian host preferences, under the assumption that the host profile of monkeys or apes is the most similar to that of Homo; (ii) preliminary adaptation to the domestic environment; (iii) exploitation of anthropogenic features of the environment as ecological markers of the most suitable habitat.[25] Mosquitoes exhibit a wide range of host preference, varying from reptiles to birds to mammals, and sometimes leading to specialized behaviour such as the anthropophilic species. From the malariological aspect, the variation in host preference can be complicated because within anopheline species complexes the host preference can be highly divergent. For example, the *Anopheles gambiae* complex consists of seven species,[26] of which only *Anopheles gambiae sensu stricto* is highly anthropophilic. *An. arabiensis* can at times feed preferentially on humans, but is behaviourally distinctly different from *An. gambiae s.s.* with a greater tendency to feed on other mammals as well.[17,27] This difference has been shown to be mediated by olfactory behaviour, *An. arabiensis* responding more strongly to carbon dioxide and less to human-specific emanations.[19] A similar phenomenon is present in the *An. fluviatilis* and *An. funestus* species complexes, where only one species each has a very strong degree of anthropophily.[18,28,29] Thus, closely related sibling species sharing the same ecological niche can exhibit widely different host preferences. As a consequence, their role as malaria vectors is also likely to be different. More specifically, in the proposed strategy of release of transgenic mosquitoes for malaria control[30] it is possible that the target species may be replaced by an incompetent mosquito, but ignorance of the other sympatric sibling species and their potential role as malaria vectors may result in a continuation of malaria transmission, albeit with reduced intensity, just as in the case of an incomplete introduced refractoriness.[31] Alternatively, the vector competence of the less suitable vector species may be enhanced by parasite-induced behavioural changes, for example by enhanced attractiveness of *Plasmodium* carriers[32] or by repetitive biting of mosquitoes carrying infectious sporozoites.[33] For this reason, the bionomics and behaviours of all potential malaria vectors in the target area need to be considered when planning a GMO approach for malaria control.

Host Abundance and Vector Behaviour

The anthropophilic malaria vectors have developed a strong association with their human hosts. In uninhabited regions and nature reserves that are situated in habitats suitable for these vectors, these species are absent. For instance, the Kruger National Park in South Africa is devoid of *An. gambiae s.s.*, while the sibling species *An. arabiensis* and *An. quadriannulatus* are widely present, feeding on the abundant wildlife.[34,35] In lowland rainforests *An. gambiae s.s.* is mostly found near human settlements, being absent in remote forests presumably due to lack of suitable hosts (M. Coluzzi, personal communication). Because humans provide the principle food source for the anthropophilic anopheline species, the transmission of human malaria parasites between humans is reinforced by the specialised feeding habits of the vectors. The density of the human population is rarely considered a factor that inhibits malaria transmission. It is not known how many mosquitoes can feed on one human host, but there is no evidence of density dependence in the population regulation of the African malaria vectors.[36] Estimates of anopheline numbers in an African village suggest that it was not the number of humans that determined the mosquito abundance in the village.[37,38] It has been suggested that zooprophylaxis might be a means for diverting mosquitoes to alternative hosts and thus reducing the human biting rate. Although this idea has been shown to work in Asia,[39,40] the African vectors cannot be sufficiently diverted to serve as effective tool for malaria control.[41]

For the vectorial capacity, however, human density is an important parameter because the human biting rate (ma) is determined by both the mosquito density and the human population density. Thus a high mosquito density with low human abundance may result in higher vectorial capacity than in a situation with median or high human abundance.[10] At present, only in urban settings with a high human density per km^2, can the figure of human density cause for sufficient dilution to affect the vectorial capacity negatively.[42]

Synchronization with the Host Habits

The synanthropomorphic anopheline mosquitoes have not only adjusted to the human environment, having developed endophagic and endophilic traits, but they have also adopted a feeding habit convergent with times when the host exhibits the least defensive responses. These anophelines blood feed between midnight and sunrise, a time when the host is usually asleep.[21,43] This enables the mosquitoes to complete their blood meals undisturbed, as during sleep the host defensive responses are likely to be small and ineffective. It is perhaps for these reasons that the use of insecticide-impregnated bed nets has been highly successful in Africa, at least in those areas where such nets have been introduced (many areas have not yet been given access to such nets), because it prevents the mosquitoes from biting when the hosts are not available, being protected by a physical barrier.[44]

Other Factors Affecting Vector-Host Contact

The development of the malaria parasite in the mosquito vector requires 10-14 days under tropical conditions. During this time, the insect will pass several gonotrophic cycles. Each cycle is initiated by a blood meal, after which the insect enters a resting stage in which its behaviour is significantly modified, with no response to host odours.[45] The suppression of host-responsive behaviour during this time serves to enhance the completion of egg maturation at a time when the insect should be left undisturbed. The traditional African mud house offers an ideal environment for this purpose, providing a dark and relatively moist environment. Malaria vectors with an opportunistic feeding preference tend to spend less time indoors, and complete the gonotrophic cycle elsewhere, where they are more exposed to environmental extremes.

Implications of Vector Behaviour for Malaria Transmission

The behaviours discussed above all contribute to enhance the transmission of malaria parasites, and it has been shown that those mosquito species with strong anthropophilic habits are highly effective malaria vectors. When considering effective intervention strategies for interruption of malaria transmission using GMO techniques, several behavioural aspects can be considered.

About ten years ago, Curtis[46] proposed that malaria vectors could be rendered zoophilic through manipulation of their genome by introgressing genes for zoophily between closely related species like the sibling members *An. quadriannulatus* and *An. gambiae s.s.* of the *gambiae* complex. The host preference is a genetic trait that may be modified, depending on the intensity of malaria transmission. In India, much of the malaria transmission is caused by *An. culicifacies*, a complex of sibling mosquitoes with mostly zoophilic species. Many of these mosquitoes bite outdoors and rest in cattle sheds. In spite of this behaviour, malaria is widespread in India, and only indoor spraying or the use of insecticide-impregnated bed nets have shown to reduce transmission effectively.[47] *Anopheles darlingi* is an important vector in South America. This species, too, is zoophilic, but can at times become associated with human settlements where it can efficiently transmit due to its high biting densities.[48,49] However, the force of malaria transmission in regions where the main vectors are mainly zoophilic is generally much lower than where vectors are highly anthropophilic, and reduced entomological inoculation rates increase the likelihood of good impact on epidemiological parameters such as malaria morbidity and mortality by traditional vector control methods. By contrast, the two most important malaria vectors in Africa, *An. gambiae s.s.* and *An. funestus*, are highly anthropophilic, endophagic and endophilic. Current control methods based on insecticide-impregnated materials where these anthropophilic vectors are present have usually had a significant impact on malaria mortality, but generally much less spectacular results on malaria morbidity.[44] In the African continent, the force of transmission is too high to achieve its interruption, or for endemicity to be destabilized.[50] It can be inferred, therefore, that for a genetic strategy based on manipulation of anthropophily to be successful, the level of penetration of the induced zoophilic trait must be complete, otherwise transmission will not be interrupted solely by partial zoophily.

It is worth distinguishing between obligate and facultative zoophily (S. Torr, C. Costantini and G. Gibson, unpublished data). Among the constraints posed by the maintenance or residual anthropophily in a facultative zoophilic vector, is the general trend for urban malaria to become the predominant epidemiological facies of the disease in Africa during the next century.[51] In the urban environment, the lack of nonhuman hosts favours human-vector contact by disallowing the normal expression of the zoophilic tendencies of the vector. Cases are known of malaria resurgence following the disappearance of the main nonhuman hosts of zoophilic vectors. In the Guyana, *An. aquasalis*, a mostly zoophilic species, shifted to biting humans and caused a malaria epidemic in Georgetown after the replacement of its main host, buffaloes used in the culture of rice, with mechanical equipment.[52] Nevertheless, integrated vector control management with existing technologies can greatly benefit from a population of vectors whose degree of anthropophily is less. Examples of successful vector control with zoophilic vectors are described in.[39,53,54]

Behavioural Genetics of Vectors

The biological basis of animal behaviour is well established: behaviour is often species-specific, it can be reproduced or altered in successive generations, it can be changed in response to alterations in biological structures or processes, and it has an evolutionary history that can potentially be traced in the genome of related organisms. The debate of the relative importance of nature vs. nurture in the ontogeny of behavioural repertoires has animated the early days of ethology when this science was still a novel scientific discipline. Nowadays, the genetic bases of behaviour cannot be denied, and the challenge for scientists in the post-genomic era is to find and disentangle the complex interaction between genes and environment at several levels of organismal organization, i.e., from the molecular interaction between stimuli and their receptors to the integration of an individual's behaviour in populations and ecosystems.

Evidences for a genetic basis of host preference are provided by three sources of information: selection experiments association between chromosomal polymorphism and feeding behaviour[55] and indirect evidence from behavioural bioassays in standardized environments[56] (H.V. Jamet (Pates), W. Takken and C.F. Curtis, unpublished data). Host preferences in malaria vectors have been shown to be already expressed early on in the behavioural sequence leading a host-seeking mosquito to its preferred host, when olfactory responses to host volatiles play a key role in the behavioural repertoire of the foraging mosquito.[57] The suitability of a host is therefore 'judged' by the profile of odorants emitted by the host. Alteration of the perceived host profile can result in the nonacceptance of the host by the questing mosquito. By manipulating the perception abilities of mosquitoes for key host volatiles, it might be possible to alter their expression of host preference. Thus, genes coding for key receptor molecules (e.g., odorant binding proteins), or promoters of receptor sensitivity are candidate targets for genetic manipulation of host preference.[58]

Genetic Manipulation for Behavioural Change?

Strongly anthropophilic mosquitoes are considered good disease vectors because of the close association with the human host (see above). For this reason, classical methods of vector control have been directed to either vector killing, for example with residual insecticides on resting sites, or prevention of mosquito bites by placing the human host under a bed net. If the bed nets are impregnated with insecticides, such nets may also result in killing mosquitoes that land on the net, although this method does not affect the entire mosquito population.[59]

Mosquito species with a more opportunistic taste for blood will be present in higher densities compared to anthropophilic species in order to cause a similar degree of transmission intensity as their anthropophilic cousins. Manipulation of the host-preference trait in malaria mosquitoes could render them less anthropophilic or even completely zoophilic, as many of the non-malaria vectors are. For instance, in tropical Africa *An. coustani* and *An. ziemanni* are both very common animal biters, occurring in high densities. Yet, these species have never been considered a vector because of their zoophilic nature. The publication of the genome of *An. gambiae s.s.* and the recent discovery of *An. gambiae* specific olfactory receptor genes[58,60] suggest that it might be possible to manipulate the odour recognition of this mosquito so that the anthropophilic trait is modified or even made extinct. It is not to be expected that mosquitoes that have thus been manipulated, will revert to anthropophilic behaviour because there are usually more animal feeds available than those on human. It is also likely that changes in the local ecosystem will render the survival chance of anopheline mosquitoes less favourable, leading to enhanced mortality or reduced adult population density. Small changes in human biting habits (parameter a, Box 1) and mosquito survival (parameter p; Box 1) can have a large impact on the vectorial capacity, thus effectively contributing to malaria reduction. These proposed changes will be less dependent on the use of genetically modified mosquitoes and therefore may be more acceptable for environmental and sociological reasons.[4] It is even conceivable that behavioural modifications can be achieved by classical selection and hybridization.[46] The factors that drive the ecology of vector behaviour and population dynamics are still poorly understood, and should be more fully explored to exploit these characters for malaria control.

Although the genome of *An. gambiae* has been identified, most of the genes that control the insect's behaviour and physiology need to be discovered. Until such information becomes available, the potential use of GMO technology other than that based on modification of vector competence,[61] remains speculative. Even then, the evolutionary forces that have resulted in the current genetic traits of the mosquito are likely to kick into higher gear to counteract the intrusion of new genetic material. For this reason we argue that ecological studies on this important group of insects, in their native habitat, should be increased to better understand the often unpredictable behaviours of entire vector populations in response to their environment.

References

1. Aultman KS, Beaty BJ, Walker ED. Genetically manipulated vectors of human disease: A practical overview. Trends Parasitol 2001; 17(11):507-509.
2. Christophides GK. Transgenic mosquitoes and malaria transmission. Cell Microbiol 2005; 7(3):325-333.
3. Holt RA, Subramanian GM, Halpern A et al. The genome sequence of the malaria mosquito Anopheles gambiae. Science 2002; 298(5591):129-149.
4. Scott TW, Takken W, Knols BGJ et al. The ecology of genetically modified mosquitoes. Science 2002; 298(5591):117-119.
5. Boëte C. Malaria parasites in mosquitoes: Laboratory models, evolutionary temptation and the real world. Trends Parasitol 2005; 21:445-447.
6. Schneider D. Using Drosophila as a model insect. Nat Rev Genet 2000; 1(3):218-226.
7. Robinson AS, Franz G, Atkinson PW. Insect transgenesis and its potential role in agriculture and human health. Insect Biochem Molec 2004; 34(2):113-120.
8. Takken W, Lindsay SW. Factors affecting the vectorial competence of Anopheles gambiae: A question of scale. In: Takken W, Scott TW, eds. Ecological aspects for application of genetically modified mosquitoes. Vol 2. Dordrecht, The Netherlands: Kluwer Academic Publishers, 2003:75-90.
9. MacDonald G. Epidemiological basis of malaria control. Bull World Health Organ 1956; 15:613-626.
10. Garrett-Jones C. Prognosis for interruption of malaria transmission through assessment of mosquito's vectorial capacity. Nature 1964; 204:1173-1175.
11. Gilles HM, Warrell DA. Bruce-Chwatt's essential malariology. 3rd ed. London: Edward Arnold, 1993.
12. Christophides GK, Vlachou D, Kafatos FC. Comparative and functional genomics of the innate immune system in the malaria vector Anopheles gambiae. Immunol Rev 2004; 198(1):127-148.
13. Trung HD, Bortel WV, Sochantha T et al. Behavioural heterogeneity of Anopheles species in ecologically different localities in Southeast Asia: A challenge for vector control. Trop Med Int Health 2005; 10(3):251-262.
14. Clements AN. The Biology of Mosquitoes. Vol I. London: Chapman and Hall, 1992.
15. Tempelis CH. Host preferences of mosquitoes. Paper presented at: Thirty-eight Annual Conference of the California Mosquito Control Association Inc., 1970.
16. Fontenille D, Simard F. Unravelling complexities in human malaria transmission dynamics in Africa through a comprehensive knowledge of vector populations. Comp Immunol Microbiol 2004; 27(5):357-375.
17. White GB. Anopheles gambiae complex and disease transmission in Africa. Trans R Soc Trop Med Hyg 1974; 68:278-299.
18. Nanda N, Joshi H, Subbarao SK et al. Anopheles fluviatilis complex: Host feeding patterns of species S, T and U. J Am Mosq Control Assoc 1996; 12(1):147-149.

19. Costantini C, Gibson G, Sagnon N et al. Mosquito responses to carbon dioxide in a West African Sudan savanna village. Med Vet Entomol 1996; 10:220-227.
20. Costantini C, Sagnon N, Della Torre A et al. Odor-mediated host preferences of West-African mosquitoes, with particular reference to malaria vectors. Am J Trop Med Hyg 1998; 58(1):56-63.
21. Haddow AJ. The mosquitoes of Bwamba County, Uganda II.- Biting activity with special reference to the influence of microclimate. Bull Entmol Res 1946; 36:33-73.
22. Maxwell CA, Wakibara J, Tho S et al. Malaria-infective biting at different hours of the night. Med Vet Entomol 1998; 12:325-327.
23. Lehane MJ. Biology of blood-sucking insects. Andover: Chapman and Hall, 1991.
24. Diatta M, Spiegel A, Lochouarn L et al. Similar feeding preferences of Anopheles gambiae and An. arabiensis in Senegal. Trans R Soc Trop Med Hyg 1998; 92:270-272.
25. Coluzzi M, Sabatini APV, Deco MAD. Chromosomal differentiation and adaptation to human environments in the Anopheles gambiae complex. Trans R Soc Trop Med Hyg 1979; 73(5):483-497.
26. della Torre A, Costantini C, Besansky NJ et al. Speciation within Anopheles gambiae - The glass is half full. Science 2002; 298:115-117.
27. Garrett-Jones C, Boreham PFL, Pant CP. Feeding habits of anophelines (Diptera: Culicidae) in 1971-78, with reference to the human blood index: A review. Bull Entomol Res 1980; 70:165-185.
28. Costantini C, Sagnon N, Ilboudo-Sanogo E et al. Chromosomal and bionomic heterogeneities suggest incipient speciation in Anopheles funestus from Burkina Faso. Parassitologia 1999; 41(4):595-611.
29. Lochouarn L, Dia I, Boccolini D et al. Bionomical and cytogenetic heterogeneities of Anopheles funestus in Senegal. Trans R Soc Trop Med Hyg 1998; 92:607-612.
30. Collins FH, Kamau L, Ranson HA et al. Molecular entomology and prospects for malaria control. Bull World Health Organ 2000; 78(12):1412-1423.
31. Boëte C, Koella JC. A theoretical approach to predicting the success of genetic manipulation of malaria mosquitoes in malaria control. Malaria J 2002; 1:7.
32. Lacroix R, Mukabana WR, Gouagna LC et al. Malaria infection increases attractiveness of humans to mosquitoes. PLoS Biol 2005; 3(9):e298.
33. Koella JC, Sorensen FL, Anderson RA. The malaria parasite, Plasmodium falciparum, increases the frequency of multiple feeding of its mosquito vector, Anopheles gambiae. Proc Royal Soc London Ser B 1998; 265:763-768.
34. Coetzee M, Hunt RH, Braack LEO et al. Distribution of mosquitoes belonging to the Anopheles gambiae complex, including malaria vectors, south of latitute 15^S. South African J Sci 1993; 89:227-231.
35. Braack LEO, Coetzee M, Hunt RH et al. Biting pattern and host-seeking behavior of Anopheles arabiensis (Diptera: Culicidae) in Northeastern South Africa. J Med Entomol 1994; 31(3):333-339.
36. Charlwood JD, Smith T, Kihonda J et al. Density independent feeding success of malaria vectors (Diptera: Culucidae) in Tanzania. Bull Entomol Res 1995; 85:29-35.
37. Taylor CE, Touré YT, Coluzzi M et al. Effective population size and persistance of Anopheles arabiensis during the dry season in West Africa. Med Vet Entomol 1993; 7:351-357.
38. Touré YT, Dolo G, Petrarca V et al. Mark-release-recapture experiments with Anopheles gambiae s.l. in Banambani Village, Mali, to determine population size and structure. Med Vet Entomol 1998; 12:74-83.
39. Kirnowordoyo S, Supalin. Zooprophylaxis as a useful tool for control of A. aconitus transmitted malaria in Central Java, Indonesia. J Com Dis 1986; 18:90-94.
40. Rowland M, Durrani N, Kenward M et al. Control of malaria in Pakistan by applying deltametrhin insecticide to cattle: A community-randomised trial. Lancet 2001; 357:1837-1841.
41. Bogh C, Clarke SE, Walraven GEL et al. Zooprophylaxis, artefact or reality? A paired-cohort study of the effect of passive zooprophylaxis on malaria in the Gambia. Trans R Soc Trop Med Hyg 2002; 96:593-596.
42. Hay SI, Guerra CA, Tatem AJ et al. Urbanization, malaria transmission and disease burden in Africa. Nat Rev Microbiol 2005; 3(1):81-90.
43. Lindsay SW, Adiamah JH, Armstrong JRM. The effect of permethrin-impregnated bednets on house entry by mosquitoes (Diptera: Culicidae) in the Gambia. Bull Entomol Res 1992; 82:49-55.
44. Lengeler C. Insecticide-treated bed nets and curtains for preventing malaria. Cochrane Database Syst Rev 2004; (2):CD000363.
45. Takken W, Loon JJA van, Adam W. Inhibition of host-seeking response and olfactory responsiveness in Anopheles gambiae following blood feeding. J Insect Physiol 2001; 47:303-310.
46. Curtis CF, Pates HV, Takken W et al. Biological problems with the replacement of a vector population by Plasmodium-refractory mosquitoes. Parassitologia 1999; 41:479-481.

47. Bhatia MR, Fox-Rushby J, Mills A. Cost-effectiveness of malaria control interventions when malaria mortality is low: Insecticide-treated nets versus in-house residual spraying in India. Soc Sci Med 2004; 59(3):525-539.

48. Rozendaal JA. Observations on the biology and behaviour of Anophelines in the Suriname rainforest with special reference to Anopheles darlingi Root. Cah ORSTOM sér Ent méd Parasitol 1987; (1):33-43.

49. de Oliveira-Ferreira J, Lourenci-de-Oliveira R, Deane LM et al. Feeding preference of Anopheles darlingi in malaria endemic areas of Rondonia State - Northwestern Brazil. Mem Inst Oswaldo Cruz 1992; 87(4):601-602.

50. Touré YT, Coluzzi M. The challenges of doing more against malaria, particularly in Africa. Bull World Health Organ 2000; 78(12):1376.

51. Robert V, Macintyre K, Keating J et al. Malaria transmission in urban sub-saharan Africa. Am J Trop Med Hyg 2003; 68:169-176.

52. Giglioli G. Ecological change as a factor in renewed malaria transmission in an eradicated area. A localized outbreak of A. aquasalis-transmitted malaria on the Demerara river estuary, British Guiana, in the fifteenth year of A. darlingi and malaria eradication. Bull World Health Organ 1963; 29:131-145.

53. Kawaguchi I, Sasaki A, Mogi M. Combining zooprophylaxis and insecticide spraying: A malaria-control strategy limiting the development of insecticide resistance in vector mosquitoes. Proc Biol Sci 2004; 271(1536):301-309.

54. Saul A. Zooprophylaxis or Zoopotentiation: The outcome of introducing animals on vector transmission is highly dependent on the mosquito mortality while searching. Malaria J 2003; 2:NIL_3-NIL_20.

55. Coluzzi M, Sabatini A, Petrarca V et al. Chromosomal differentiation and adaptation to human environments in the Anopheles gambiae complex. Trans R Soc Trop Med Hyg 1979; 73(5):483-497.

56. Gillies MT. The role of secondary vectors of Malaria in North-East Tanganyika. Trans R Soc Trop Med Hyg 1964; 58:154-158.

57. Foster WA, Takken W. Nectar-related vs. human-related volatiles: Behavioural response and choice by female and male Anopheles gambiae (Diptera: Culicidae) between emergence and first feeding. Bull Entomol Res 2004; 94(2):145-157.

58. Zwiebel LJ, Takken W. Olfactory regulation of mosquito-host interactions. Insect Biochem Molec 2004; 34(7):645-652.

59. Hawley WA, Phillips-Howard PA, ter Kuile FO et al. Community-wide effects of permethrin-treated bed nets on child mortality and malaria morbidity in western Kenya. Am J Trop Med Hyg 2003; 68:121-127.

60. Merrill CE, Pitts RJ, Zwiebel LJ. Molecular characterization of arrestin family members in the malaria vector mosquito, Anopheles gambiae. Insect Mol Biol 2003; 12(6):641-650.

61. Riehle MA, Srinivasan P, Moreira CK et al. Towards genetic manipulation of wild mosquito populations to combat malaria: Advances and challenges. J Exp Biol 2003; 206(Pt 21):3809-3816.

CHAPTER 12

Genetically-Modified Mosquitoes for Malaria Control:
Requirements to Be Considered before Field Releases

Yeya T. Touré and Bart G.J. Knols*

Abstract

The technical feasibility of the development of transgenic mosquitoes highly refractory to (rodent) malaria parasites has been demonstrated in the laboratory. Following this proof of principle, genetic control of vectors could have an important role to play in the interruption of transmission of human malarias, if the main developmental and implementation challenges are adequately addressed. These include the establishment of a proof of efficacy and safety for humans and the environment in carefully controlled and contained environments. Prior approval by authorized biosafety, regulatory, and ethical review bodies needs to be obtained before experimental releases. In addition, there is the need to ensure the public and the media that this process is desirable, feasible and can be accomplished safely. Moreover, an appropriate implementation and capacity building plan would increase the chances of making this approach a control method applicable for public health purposes. Analysis of current and anticipated future views of a variety of critical stakeholders enables the provision of a framework that facilitates the transition of research findings from the laboratory to the field. A coordinating mechanism to closely monitor and guide this transition process will be instrumental in furthering developments to fully evaluate the public health potential of this approach.

Introduction

Mortality and morbidity from malaria remain high despite a global commitment to its control.[1,2] Several reasons including poor implementation of interventions, development and spread of resistance of parasites to antimalarial drugs and of vectors to insecticides, besides insufficient human resources, have been shown to contribute to this high disease burden.[3]

Malaria prevention relies mainly on vector control/personal protection measures and chemoprophylaxis. The vector control methods mostly used are indoor residual spraying of insecticides and the use of insecticide-treated bednets or curtains. In many settings, these strategies face implementation and sustainability problems.[4]

As the actual control methods and strategies are being applied for malaria prevention, it is crucial to continuously look forward for their improvement and for the development of new and innovative ones. In this regard, on the basis of the progress made in molecular biology and in biotechnology (e.g., genetic modification of *Drosophila*[5]), it was thought that the time is ripe

*Corresponding Author: Bart G.J. Knols—International Atomic Energy Agency (IAEA), Agency's Laboratories in Seibersdorf, Seibersdorf, Austria. Email: B.Knols@IAEA.org

Genetically Modified Mosquitoes for Malaria Control, edited by Christophe Boëte.
©2006 Landes Bioscience.

to explore the feasibility of interrupting malaria parasite transmission through the genetic modification of its vectors.[6]

The technical feasibility of the development of transgenic mosquitoes with impaired ability to transmit (rodent) malaria parasites has been demonstrated in the laboratory. *Anopheles gambiae* (Giles) has been genetically modified and *Anopheles stephensi* (Liston) was made highly refractory to *Plasmodium berghei* growth and transmission.[8] These achievements open an avenue for potential contribution of genetic modification of vectors to malaria control. However, challenges about the complete development of the method, its implementation and public concerns remain to be addressed and should provide an evidence base for policy decision, which would facilitate this approach to become a tool for use in public health for malaria control. Consequently, lessons are to be learned from the genetically-modified food debate[9] and previous genetic control trials of vectors in El Salvador[10] and India.[11]

Here we analyse the issues and challenges to be considered before field releases and highlight how these could be addressed.

Developmental Challenges

The biotechnological challenges about the development of the control method include issues such as developing and evaluating appropriate effector gene constructs, devising and testing suitable gene driving systems, assessing the spread of the foreign genes and fitness impacts on mosquitoes. In addition, a much more challenging undertaking is to provide a proof of efficacy and safety for humans and the environment of the method under laboratory and semi-field conditions.

A major concern about the development of transgenic mosquitoes for malaria control is represented by unexpected biological changes, which could affect their transmission capabilities. The transgenic mosquitoes must remain comparable to their field counterparts with the only difference that they would have gained the ability to prevent malarial infections. But in addition, they would not be expected to be able to transmit other pathogens such as filarial worms or (arbo)viruses.

In order to address these issues and reassure the public, studies need to be conducted on the efficacy, bio-safety and risk/benefit evaluation through long-term efforts to clarify the scientific uncertainties under different experimental conditions.[12-14] This activity is best undertaken in partnership between researchers from developed countries and from disease-endemic countries.[15] In addition, the participation of the public and the media is also highly recommended. A sound basis for collection of data on vector biology, ecology, behaviour and genetics addressing efficacy and safety in the field would also need to be provided.[16] Moreover, there is the need to develop criteria to identify and prepare the field sites earmarked for anticipated releases. There should be prior environmental and health studies for site selection, and based on these data the most appropriate sites should be chosen. Guidelines and principles would need to be developed on the design and performance of efficacy and minimum risk field research. Criteria and test methods for environmental monitoring are also required.

An appropriate safety assessment and management would represent a sound basis for policy decision. Its conduct would need the identification of scientific principles and practices for undertaking safe laboratory experiments and field trials with genetically-modified mosquitoes following Good Developmental Practices (GDP).[18-19] In addition, it needs to include mechanisms to provide the public with information about the bio-safety assessment results and ensure the information reaches the communities and decision-making bodies.

Public Concerns and Implementation Challenges

Important achievements have been made in biotechnologies, which could potentially be used for malaria control. However, the public in general and particularly in disease-endemic countries (DECs) is not sufficiently informed about the potential and the process of development of new technologies such as genetically-modified mosquitoes. This information gap affects the capability of the public and policy makers alike to fully judge the efficacy and safety of genetically-modified mosquitoes for humans and the environment and to make informed decisions about their implementation for malaria control. Currently, the public in disease-endemic countries, as future end users of this approach, are insufficiently involved in the adoption process with the imminent risk that it may ultimately use its power to reject it. Consequently, mechanisms need to be developed for information exchange, for addressing public concerns and for using standardized regulations world wide in a coordinated manner.[20]

As a first step, it is necessary to set up mechanisms for improved communication, provision of adequate means for information dissemination and collaboration between the researchers, the public and the media. An adequate translation of scientific knowledge to the public and the media is highly desirable. The information should be openly provided as broadly as possible in a reciprocal process.[14] This procedure would ensure an appropriate flow of information exchange and feedback, which would result in raising public awareness, addressing concerns about potential environmental and human health risks and building public confidence in the scientific results. It will also provide means to the public to be sufficiently knowledgeable to make informed decisions about the merits of deploying such programmes in their communities.

Another necessary step would be to bring all parties together on common ground that can lead to objective, scientific, legal, ethical and social-based decisions by policy makers, whilst bearing in mind that most people may not trust the scientific efficacy and risk analyses.

Ethical, legal and social issues (ELSI) of the use of genetically-modified mosquitoes need to be fully addressed in order for the community at large to adhere to the principle of their potential implementation. For this to be achieved, it would be necessary to integrate with the scientific studies those legal, ethical and social factors that are relevant to the use of genetically-modified mosquitoes and ensure that all parties with legitimate concerns have mechanisms for including their input into the proposed genetic control programmes. In addition, it is necessary to engage the end-users in the choice of sites and plans for deployment, in clear and legally-appropriate concepts of informed consent, in promoting an understanding of the real measures of success for the programmes. Moreover, consent should be obtained from the communities involved and the mechanisms to obtain individual and group consent need to be specifically developed for public health interventions based on genetic vector control concepts. The data should be made open to all so that it can benefit from global expertise and develop an international consensus.[14]

The safety assessment plan during the implementation phase would need to establish the development of science-based evidence for policy and procedures for the assessment and management of potential risks. Moreover, it will aim at minimizing the potential adverse human and environmental consequences. More specifically, it will have to anticipate detrimental effects that might follow the release of genetically-modified mosquitoes during experimentation. It would also need to design monitoring systems for the early detection and evaluation of adverse outcomes, and plan interventions strategies, so that new information can be gathered and interpreted to avert and if necessary remedy adverse health or environmental effects.[21] Guidelines are needed for assessing dispersal, contingency measures and site rehabilitation.

The approval of the implementation plan of genetically-modified vectors as a control strategy should be based on a proof of efficacy and safety properly established and approved by authorized bio-safety regulatory and ethical review bodies before any experimental release.[12-14] The information necessary for legal and regulatory approvals should be gathered and should include a complete documentation for bio-safety and ethical review. The requirements necessary for national and local authorities' approvals should also be addressed. The development of guidelines and regulatory procedures would help the researchers and the countries to define a common ground to deal with the issues about the processes for potential approval of the implementation of genetically-modified mosquito-based vector control.

Capacity and Partnership Building and Coordination Mechanisms

In order for DECs to be fully involved in undertaking the evaluation and potential implementation of genetically-modified mosquitoes as a control tool, there is the need to enhance their capacity, build partnership and set up a coordination mechanism. Investigators would need to receive funding for research and training to undertake activities for collecting data on the vector biology, ecology, genetics and behavioural ecology of the vectors. They would need to be trained for monitoring and managing safety procedures for human health and for the environment, for evaluating risk/benefits and for managing regulatory and ethical principles. DECs need to be helped to create and manage institutional/national bio-safety and ethical review boards. There would also be the need for promoting South-South and North-South research collaboration based on well-defined ethical and scientific standards.[22]

The complexity of issues related to genetically-modified mosquito development and implementation requires a multi-disciplinary effort, which would need an international coordinating board. It will focus on the broader dissemination of scientific progress to stakeholders, the facilitation of collaborative efforts and partnership strengthening within and beyond the scientific community and the mobilization of financial resources for the funding of developmental and implementation plans.

Conclusion

The technical feasibility of genetically-modified mosquitoes highly refractory to rodent malaria parasites has been demonstrated under laboratory conditions. However, there are developmental and implementation challenges to be addressed before this method can be used as a public health tool. The public perception about the use of genetically-modified mosquitoes varies from fear and refusal to hope and concerns, because of uncertainties and information gap. For this reason, there is a need for careful and thorough assessment of efficacy and safety and addressing properly the public concerns before possible implementation. There is the need to provide adequate means for information dissemination, communication and collaboration between the researchers, the public and the media such to raise awareness and address concerns about possible environmental and human health risks. Ethical, legal and social issues of the use of genetically-modified mosquitoes need to be fully addressed in order for the community at large to appreciate value and endorse the principle of their potential implementation. The implementation of GMM needs public acceptance and appropriate plans for monitoring and managing potential risks over time under well defined regulatory, capacity building and coordination mechanisms.

A meeting, jointly-organised by WHO/TDR, IAEA, NIAID and Frontis (Wageningen University, The Netherlands), held in Nairobi in July 2004, focused on the above issues and developed the following seven recommendations:[15]

1. Genetic modification of insects could be used to control vector-borne diseases, yet will depend on solving several critical components of the approach, i.e.,
   • Optimising currently available transformation systems;
   • Identification of additional endogenous and/or artificial effector genes, conditional lethals and novel phenotypes;
   • Identification of tissue-specific promoters for such systems.
2. Develop techniques for driving effector genes that interfere with disease transmission into wild insect populations, i.e.
   • Research on the genetic stability of effector and drive mechanisms and their associated fitness costs.
3. Studies on vector field populations with respect to potential future releases of genetically-modified mosquitoes, i.e.,
   • Understanding male mosquito biology and particularly the factors affecting mating and competitiveness under field conditions;

- Development of models to define threshold levels in terms of system efficacy in order to attain maximum epidemiological impact in both spatial and temporal dimensions;
- Characterisation of field sites and field populations that should include the establishment of relationships between transmission intensity and disease outcome.

4. Development of processes dealing with the ethical, legal and social issues (ELSI) of the use of genetically-modified mosquitoes, i.e.,
   - Any genetically-modified mosquito approach must result in a predictable and positive public health outcome;
   - Development of guidelines and principles for minimum-risk field research that includes environmental risk management.

5. Enhanced involvement of scientists and institutes in DECs. As frontline stakeholders in this endeavour, the roles and responsibilities of DEC scientists should increase and be based on equitable partnership development.

6. Inclusion of genetically-modified mosquitoes in disease control programmes, i.e., research that focuses on the inclusion of genetically-modified mosquito approaches within the broader framework of malaria vector control. Integration of genetically-modified mosquitoes into integrated vector management (IVM) programmes will need to be considered.

7. Coordinating and follow-up mechanism for genetically-modified mosquito research and implementation. Considering the complexity and multi-disciplinary nature of this endeavour, the establishment of a coordinating board was recommended. This board should oversee research developments and drive the broader dissemination of research results to stakeholders and facilitate collaborative research and partnership strengthening within and beyond the scientific community.

References

1. Snow RW, Guerra CA, Noor AM et al. The global distribution of clinical episodes of Plasmodium falciparum malaria. Nature 2005; 434:214-217.
2. Malaria: A major cause of child death and poverty in Africa. UNICEF/Roll Back Malaria 2004; 18.
3. Breman JG, Alilio M, Mills A. Conquering the intolerable burden of malaria: What's new, what's needed. Am J Trop Med Hyg 2004; 71(suppl 2):1-16.
4. World Malaria Report 2005. Roll Back Malaria/World Health Organization/UNICEF 2005, (Publ. WHO/HTM/MAL/2005.1102).
5. Spradling AC, Rubin GM. Transposition of cloned P elements into Drosophila germ line chromosomes. Science 1982; 218:341-347.
6. Anonymous. Prospects for malaria control by genetic manipulation of its vectors (TDR/BCV/MAL-ENT/91.3). Geneva: World Health Organization, 1991.
7. Grossman GL, Rafferty CS, Clayton JR et al. Germline transformation of the malaria vector, Anopheles gambiae, with the piggybac transposable element. Insect Mol Biol 2001; 10:597-604.
8. Ito J, Ghosh A, Moreira LA et al. Transgenic anopheline mosquitoes impaired in transmission of a malaria parasite. Nature 2002; 417:452-455.
9. Knols BGJ, Dicke M. Bt crop risk assessment in the Netherlands. Nat Biotech 2003; 21:973-974.
10. Lofgren CS, Dame DA, Breeland SG et al. Release of chemosterilized males for the control of Anopheles albimanus in El Salvador. III. Field methods and population control. Am J Trop Med Hyg 1974; 23:288-297.
11. Anonymous. WHO-supported collaborative research projects in India: The facts. WHO Chronicle 1976; 30:131-139.
12. Hoy MA. Deploying transgenic arthropods in pest management programs: Risks and realities. In: Handler AM, James AA, eds. Insect transgenesis. Methods and applications. USA: CRC Press LLC, 2000:335-367.
13. Alphey L, Beard CB, Billingsley P et al. Malaria control with genetically manipulated insect vectors. Science 2002; 298:119-121.
14. Macer D. Ethical, legal and social issues (ELSI) of genetically modified vectors in public health. TDR/STR/SEB/ST/03.1. Geneva: World Health Organization, 2003.
15. In: Knols BGJ, Louis C, eds. Bridging laboratory and field research for genetic control of disease vectors. Springer/Frontis, 2005:215.
16. Scott TW, Takken W, Knols BGJ et al. The ecology of genetically-modified mosquitoes. Science 2002; 298:117-119.

17. Knols BGJ. Identification and characterization of field sites for genetic control of disease vectors. In: Knols BGJ, Louis C, eds. Bridging laboratory and field research for genetic control of disease vectors. Springer/Frontis, 2005:Chapter 20.

18. Touré YT, Oduola AMJ, Sommerfeld J et al. Biosafety and risk assessment in the use of genetically modified mosquitoes for disease control. In: Takken W, Scott TW, eds. Ecological aspects for application of genetically modified mosquitoes. Kluwer Academic Publishers, Frontis Series Vol 2, 2003:217-222.

19. Touré YT, Oduola AMJ, Morel C. The Anopheles gambiae genome: Next steps for malaria vector control. Trends Parasitol 2004; 20:142-149.

20. Anonymous. Bugs in the system? Report prepared by the Pew Initiative on Food and Biotechnology 2004; 1-109.

21. Report of the Edmonds Institute. Manual for assessing ecological and human health effects of genetically engineered organisms. By a Scientific Working Group on Biosafety. Washington, USA: Publication of The Edmonds Institute, 1998.

22. Mshinda H, Killeen GF, Mboera LEG et al. Development of genetically-modified mosquitoes in Africa. Lancet Inf Dis 2004; 4:264-265.

CHAPTER 13

Ethics and Community Engagement for GM Insect Vector Release

Darryl Macer*

Abstract

The ethical, social and legal issues raised by the release of genetically modified insect vectors in public health need to be considered in depth at an early stage in the development of protocols to field test GM insects. This chapter also examines the use of GM technology applied to mosquitoes for malaria control in general. There is a need to engage the community and have two way communication between researchers, policy makers and local communities in order to find whether each particular community will want to have a field trial, the nature of the concerns they have, and the ways that can be designed to involve communities as partners in trials.

The Ethics of Disease Prevention

This chapter will examine the ethical issues that underlie efforts to control human disease, modify vectors, modify the environment and methods to seek community support. There is global support for the efforts to improve existing and develop new approaches for preventing, diagnosing, treating and controlling infectious diseases that cause loss of human life.[1] The ethical principle that lies behind the idea of preventing, treating and controlling disease is that human life should be protected. We can debate what are the most ethical measures for achieving these goals, including the extent to which risks to human health, damage to the environment and other living organisms, and economic costs are balanced in societies that have a range of worldviews and social structures.

Certain principles basic to resolving ethical dilemmas can help decision makers make more informed policy decisions. The principle that we should love the life given to us (self-love) implies that each person should be given autonomy (self-rule) to work out how to balance the ethical dilemmas and choices themselves. The Universal Declaration of Human Rights of 1948 specifically set as a baseline that all human beings possess equal rights, and should be given a chance to exercise their autonomy. One of the fundamental human rights is a right to health, and working towards giving every person a chance to grow up free of disease is the ethical foundation of public health. If a person does not possess some basic level of health, he/she cannot even face many of the choices commonly accepted as normal. Poverty also restricts the choices of many people,[2] especially in areas faced with infectious insect borne diseases.

Justice simply means that if we want others to recognize our autonomy, we have to recognize theirs as well. There are at least three different meanings of the concept of justice: compensatory justice - meaning that the individual, group, or community, should receive recompense

*Darryl Macer—RUSHSAP, UNESCO Bangkok, 920 Sukhumwit Road, Prakanong, Bangkok, Thailand 10110. Email: d.macer@unescobkk.org

Genetically Modified Mosquitoes for Malaria Control, edited by Christophe Boëte.
©2006 Landes Bioscience.

in return for contribution; procedural justice - meaning that the procedure by which decisions about compensation and distribution are made is impartial and includes the majority of stakeholders; and distributive justice - meaning an equitable allocation of, and access to, resources and goods.[1] There are ethical questions about how a society should represent procedural justice when there are major divisions within the society on particular issues, as we find in many countries with debates over the use of genetic engineering. The process of consensus building and reaching common ground may be preferable for many cultures rather than confrontations.

At present there is great inequality between rich and poor nations in the direction and priorities of research, and in the distribution of and access to benefits that might come from this research. Under any ethical theory, the presence of diseases that threaten the lives of not just one but more than a billion people worldwide provides a compelling need for efforts to eradicate the diseases. There is wide diversity in the risks that members of each community face from infectious diseases due to: individual genetic variation in resistance to infectious disease agents; a person's nutritional state and immediate environment; a family's economic situation with respect to providing barriers to vectors and disease; access to both preventative and therapeutic medicines. These variations can be regarded as a type of lottery. Working towards better global equity is a goal that attempts to even out the lottery that people are born into. This is ethically mandated by Rawlsian justice,[3] which argues that efforts should be made to minimize the variation in all social factors because no one knows before they are born into which situation they will be born, so everyone would wish for equal opportunity and equal exposure to risk. All should have a chance to be born and grow up in an environment free of infectious diseases, if that can be achieved.

The ethical principle of beneficence supports the development of science and medicine, and its provision to those who suffer. A universal ideal found throughout human history is that it is better to love doing good things than bad things, and to love our neighbour as ourselves.[4] Humans have used technology in efforts to make their lives easier and better for thousands of years, and the ethical principle of beneficence argues that we should continue to make life better. This ethical principle is based on the general motivation inside people to love doing good rather than harm, and may be expressed as love or compassion.[5] Efforts that work for the betterment of others in society have a universal moral mandate.

The ethical principle of non-maleficence, or do no harm, would make us reasonably cautious about premature use of a technology when the risks are not understood. Recently some have advocated a total precautionary principle for genetic engineering, which would mean that no technology with more than 0% risk should ever be attempted.[6] This has also entered the Cartegena Protocol on Biosafety, which is an International Legally Binding Agreement that regulates international movement of living modified organisms (LMOs).[7] Because no human action has 0% risk, the principles of both benefit and risk are used to assess technology and are central to any public health program.[8]

The ethical issues raised by biotechnology are commonly termed bioethics dilemmas, although when we examine the actual moral questions they may not be so novel and are often related to areas of applied ethics that were debated long before we had modern biotechnology.[9] There are several basic theories of ethics. The simplest distinction that can be made is whether they focus on consequences, actions or motives. Consequential arguments are the criteria applied to assess the ethics of biotechnology applications, i.e., whether they contribute to the greater good by improving the well-being of all. Consequential arguments state that the outcome can be used to judge whether an action was ethically correct or not. An action-based argument looks at the morality of the act itself, so that the actual action to cause harm itself is an unethical action regardless of the consequences or motives. Motive-based theories of ethics, including virtue-based ethics, judge an action based on the motivation of the action. For example, if the act was done with good intentions or not. Another separation that is used is between deontological theories, which examine the concepts of rights and duties, and teleological ones, which are based on effects and consequences. If we use the image of walking along the path of life, a teleologist tries to look where decisions lead, whereas a deontologist follows a planned direction.

The objects and subjects of ethics can be viewed in terms of ecocentric, biocentric or anthropocentric concerns. Ecocentric concerns, that value the ecosystem as a whole, are used when expressing environmental concerns. The reverence for all of life[10] can apply to the whole ecosystem or to every member of it. Biocentric thinking puts value on the individual organism, for example one tree or one animal. Anthropocentric thinking is focused on the human individual. There is a trend for more ecocentric views to be included in recent legislation, with protection of ecosystems for their own value. While it can be useful to isolate distinct issues, as will be done in this report, it is not realistic to separate human/nature and social interactions. This is because almost all of human life is a social activity, involving many relationships with people and the ecosystem. Different ethics are implied when human activity, e.g., agriculture or urbanization, attempts to dominate nature or to be in harmony with the environment.

Despite the fact that there are a variety of definitions of health, disease, disability, and what is a meaningful human life, working to alleviate disease and empower individuals to reach their potential are universal goals for the progress of humankind. The basic ethical principles of autonomy, justice, beneficence and non-maleficence can be applied to help decision-making in a range of bioethical dilemmas in medical and environmental ethics. There is some debate over whether further principles can always be derived from these over the precise terminologies in each field,[11] but the general consensus is that these four principles are fundamental in a range of cultures.[12,13] The emphasis on individuals may be questioned more in developing countries. There are also theories of ethics based on community, which argue that individuality, autonomy or rights of a person are not suited to the community structure of society.

Ethics of Genetic Manipulation

There is a long history of altering the behaviour of disease vectors so that they cannot transmit pathogens to humans.[14] Previous chapters have discussed the scientific background. Insects have also long been the targets of attention in agriculture as well as in medicine. While there are few intrinsic ethical concerns about killing insect pests, as discussed below, ecocentric approaches to ethics do raise some objections to modification of ecosystem components, and these need to be taken more seriously.

People of all cultures have developed biotechnologies as they live together with many species in the wider biological and social community. A simple definition of biotechnology is the use of living organisms (or parts of them) to provide goods or services. Over five millennia of classical plant and animal breeding have seen the emergence of agricultural societies, and modern biotechnology is built on that. Since the mid 1990s, foods produced from genetically modified organisms (GMOs) have been sold in a growing number of countries.[15] There has been fierce international debate over the environmental and human health aspects of GM foods, but no harmful effects of GM foods on human health have been shown scientifically until now.[16] There still remain doubts in some quarters though on how we could detect if there were any affects, and precaution is applied to avoid involving known allergens in GM food. There is almost no practical scientific measure available to measure the long term effects of any foodstuff because most people consume such a variety of foods and substances. There was concern over StarLink Corn that was only approved for animal consumption entering the human food chain, so that the policy was changed to only approve varieties of plants for animal consumption if they will also be approved for human consumption. The US FDA concluded an investigation, however, that despite claims of allergic reaction, there were no adverse human health reactions.

There is greater concern over the environmental impact of gene transfer in the environment, and these include concerns about cross-pollination of wild relatives of rapeseed to fears of gene transfer between maize in areas of genetic origin, such as Mexico. A number of governments have considered the issues and concerns people have raised about genetic engineering, and there is a wealth of useful material on the pros and cons in the reports and submissions made to them.[17,18] Reports have also been made by independent organizations on the ethical issues.[19] With the emergence of genomic sequencing, we now have the DNA sequence of human beings, dozens of pathogens, and some disease vectors e.g., *Anopheles gambiae.*[20,21] It is therefore not surprising that molecular entomology, the study of DNA and the proteins it encodes in insects, is emerging as a serious scientific approach for insect control.[22,23] Social factors need to be carefully considered.[1,24] While there is debate over the use of funds to combat infectious disease using genomics and biotechnology as opposed to implementing practical measures to curb vectors and pathogens in the field,[25] it is hoped that the former approach will be a major strategy in the future.[26,27] A common way to insert DNA for genetic transformation of insects is to use transposons or viruses.[28] A number of papers in this book describe the advances that are being made in this field. Most attention has been given to efforts to genetically transform insects in the laboratory, and to test their behaviour before releasing them into the environment. A mechanism that would safely spread the gene among vectors in the wild is the objective of these studies, except for the approach using sterile insects. Effector mechanisms are needed to drive the effector system into the vector population,[29] which raises more ethical issues about the safety and desirability of changing the entire vector population, and possibly related species.

The conclusions of studies of ethical issues inherent to the process of genetic engineering compared to traditional methods of animal and plant breeding, are that the only significant differences in the process are the more precise control of genetic engineering and whether the DNA involves cross-species gene transfer that does not occur in nature.[1,9,19] One of the key questions is whether there is an intrinsic value of genetic integrity at an organism and ecosystem level that humans should not change. There are some persons in some communities that place intrinsic value upon native fauna including insects, however the way that they do would require well designed research to investigate. We should also note that cross-species DNA transfer does occur in nature between all species, even of different kingdoms, and that the genomes of insects are subject to genetic flux in nature. In this sense, because the DNA change can be precisely designed, an actual targeted genetic change through genetic engineering should be safer than a natural change because it is more under control. However, issues of control are raised if a 'transposable element + allele of interest' are inserted, especially when the strategy is for wide spread of this in the wild.

Given the results of public opinion surveys that find opposition to cross species gene transfer,[30,31] if the DNA change is made using DNA within the same species entirely, then this concern can be removed. In this way of thinking there may not be any new intrinsic ethical dilemma from the modification of DNA structure in genetic engineering as it simply mimics the natural ways organisms use to change genetic structure. However, the scientific details of the targeting process, and the intentional nature (the issue of control of nature) are important for some persons. There may be a distinction between use of a naturally occurring DNA sequence that was transferred between species to the use of an artificially designed novel sequence, although from a chemical point of view these are both DNA sequences producing peptides. For GM mosquitoes the idea is more to introduce an artificial piece of DNA that will lead a peptide. This one will then render the mosquito able to kill/ block the parasite.

Mosquitoes and Animal Rights

Another concern in ethics when discussing animals is their capacity to suffer or feel pain. If insects do not feel pain or sense feelings, then the most prevalent ethical approach for animals would argue that there is nothing intrinsically wrong in manipulating them.[32] Given what we know about mosquitoes in this approach they would have no moral rights. However, if we consider the idea of making so-called vegemals, animals that do not feel pain, we are still manipulating life for human purposes without considering the interests of the animal.[33] The concern is that living organisms should not merely be treated as a means to the ends desired by humans. There are also extrinsic values placed on some animals by human society, but I do not know of any which place special value upon mosquitoes. There are biodiversity concerns about endangered animals in general, some of which are expressed in the Convention on Biological Diversity.

Another argument used in these discussions concerns the telos (purpose) of an organism. A teleological explanation describes phenomena by their design, purpose, or final cause. Teleology is the branch of moral philosophy dealing with the cause and effect of an action, the belief that there is purpose and design in nature, and consequently, with the belief in the existence of a Creator. There are concerns that the ability to alter the telos of an animal has profound implications.[34] If one believes that every organism has a purpose, then the telos is an intrinsic concern, and genetic engineering alters the telos or 'being-ness' of an organism. However, it is debatable whether changes and control through genetic engineering are significantly different from changes made by humans to animals and plants in farming and modern life. It is basically an issue of human control of nature, and there is debate over the extent to which humans should control nature.[9,35,36] If we consider this issue in a historical context, we see that humans in many affluent cultures have controlled nature in significant ways, e.g., by concrete river banks, irrigation and sanitation projects. However, especially in some developing countries, limited resources have meant that control of nature has been less. However, sociological evidence has found that a number of people object to human control of nature, regardless of whether it poses a risk.[30]

While perhaps only followers of the Jain religion in India regularly refrain from killing insects that are human pests, there are still some people who may object to killing mosquitoes. It is not known if manipulating the insects so that they would not be a human pest and would still remain a species in the ecosystem would be more acceptable to persons with these ecocentric world views than traditional methods of insect control that attempt to eradicate a whole insect population, often affecting a number of insect species. However with GM mosquitoes built to resist malaria infection the idea is more to kill the parasite inside the mosquito rather than killing the mosquito as this technique consist of replacing a vector population by a non-vector population. Moreover it is very likely that before any field test or release of GM mosquitoes, insecticides will be used before to reduce the local population to facilitate the success of the GM mosquitoes. So some killing will still be involved.

Those who subscribe to an ecocentric viewpoint might argue that the ecosystem as a whole would benefit from an intervention that left the mosquitoes in the ecological community, with the elimination of the disease-causing pathogen from the vector, if the alternative was eradication of the vector species. In this case the total number of species affected by this type of genetic modification of vectors would be significantly less than the number of species affected by use of insecticides.[1] However, there are still those who believe there should be no human modification of the ecosystem. This actually should argue that there should be no direct or planned modification of an ecosystem by humans, since human activity modifies almost all ecosystems, including those where humans are not directly a component member.

Community Engagement and Environmental Risks

The process of community engagement has several goals, as developed recently in human genetic studies.[37] It should approach a broad range of members of the communities for participation in a two way process of information exchange to share with investigators their views about the ethical, social, and cultural issues the scientific project raises for them, their immediate communities, and the broader communities and populations of which they are a part. It should provide input that may modify the disease control mechanisms and approaches that will be adopted. It should provide extensive information about the project so that the decisions of individuals about whether or not to support their community involvement would be better informed. It also will be expected to continue throughout the trials, including sharing findings from studies conducted.

There will be expected to be negligible human risks from the trials of GM insect vectors, but still consent should be considered. Firstly, let us consider environmental risks of a trial because the GM insect vectors may represent potential harm to other members of the biological community as well as other members of the human community. Globally people vary in the importance they ascribe to the environment, or parts of it. Especially in areas where more traditional world views are found, we may see greater value given to parts of the environment that are forgotten in the modern industrial mindset. We also see variations between persons in all cultures as to their images of nature and what is life.[30] Some people are willing to sacrifice themselves for the environment. Examples such as the preservation of sacred groves in India for thousands of years, even during times of severe crisis and human death,[38] show that in some cultures almost all people are willing to die rather than damage that part of the environment they cherish. This behaviour is often linked to religious beliefs in the afterlife.

A variety of potential broader ecological, environmental and health risks are associated with the release of GM organisms. Environmental risks can be considered from both anthropocentric and ecocentric-based approaches. The risks identified include the possibility of horizontal transfer of the transgene to non-target organisms, and possible disturbance of insect ecology.[19,39] There have also been concerns expressed in some cultures, e.g., New Zealand, over the need to value the native fauna and flora, which is considered by many in the Maori community to be something not to modify.[18] While human beings cannot consent for other organisms to be modified, very few persons suggest that any consent is required except for possibly sentient animals.

One of the main concerns of releasing GMOs is environmental risk.[40,41] This risk has been controlled in over 10 000 international field trials of GMOs, and in the widespread commercial growth of GM crop varieties.[42] Whilst the methods used for monitoring field trials are argued to be inadequate by those campaigning against GMOs, to date there has not been a significant adverse event from GMO release for the health of any non-target organism, including humans, in the ecosystem.[9] There are a range of concerns that have been expressed, including cross-pollination of non-GM crops with GM crops, economic dependence on the seed industry, intellectual property, for example.[19] There are also cases which have questioned the degree of control over the process that has been attained, and also which is possible. The question of so-called genetic pollution, gene flow to wild relatives, is still in a process of debate when comparing gene flow in nature, gene flow in conventional agriculture and gene flow in systems with GM varieties. Farmers also may grow seed in fields and for purposes that were not intended by the seed makers, or industry may mix food products, as seen in the entry of StarLink varieties into Tacos for human consumption. This raised concerns over how policy can be implemented and policed at the practical level.

In the year 2001, the first US field test of a genetically modified pink bollworm, a cotton pest, was conducted. It followed very soon after the development of methods to transform the bollworm.[43] This type of trial had an important consequence of better preparing regulatory systems for oversight of GMOs/LMOs, but still most countries in the world have not established detailed systems for oversight of GM insect field releases.[44] The American Committee of Medical Entomology has also produced guidelines.[45] New ethical issues about GM arthropod vectors and their symbionts and/or pathogens should be subject to extensive open discussions and forums where not only experts but all members of civil society should participate.

Any risks to the agricultural systems of rural communities also require assessment, as animal diseases transmitted by vectors are important to farming families. In addition, there may also be risks to wild animals in surrounding areas, which in some ecocentric environmental views have more intrinsic rights to be left undisturbed than farm animals.[46] This calls for broad ecological understanding of the impact, beyond public health. There is also the possibility for GM vectors to spread to areas beyond the initial expectations, which needs to be considered when planning the geographical extent of information and communication programs.

Although there have been numerous public opinion surveys on the release of different GMOs, there have been few surveys asking people their views on introducing GM vectors or pathogens for disease control. One general feature of the surveys is that GM plants are considered less threatening than GM microbes, animals and humans. In a 2003 national sample in Japan, one third thought it would be acceptable to use genetic engineering to make mosquitoes unable to be a vector for human diseases like malaria or Japanese encephalopaty, and only 16% said it would not, while half said they did not know. There was 54% approval for environmental release of mosquitoes that do not transmit human disease, which is the same as the support for release of GM disease resistant crops, with 19% disagreeing.[47]

Although knowledge is important for acceptance of biotechnology, it is not a predictor of acceptance. In surveys of scientists and the public in Japan in 1991-2000, for example, well-educated scientists were often just as sceptical of biotechnology as the general public, and shared the same types of concerns.[31] The failure of the government authorities in public health has led to higher public trust in NGOs, including environmental groups. The media has also disproportionately reported negative aspects of genetic engineering because these appeal to people,[48] while other groups in society have promoted biotechnology for commercial purposes.

Thus the late 1990s saw a dramatic drop in public support for biotechnology in every country surveyed. It is therefore important that scientific knowledge be accurately shared with all, that this process be open, and that all opponents are involved in discussion.

Issues include the ethics behind research into, and later financing of, technological products that attempt to "fix" a problem rather than invest in increasing the ecological knowledge base to "prevent" the problem. There is considerable preference for deterministic science over "softer" educational systems like flexible learning. It is clear that not all local communities will share the modern scientific world view that technical healing is better for them, so there needs to be flexibility in the approaches available to eradicate disease. In the past, paternalistic interventions were taken on the behalf of citizens; however, civil rights movements have empowered people to take these decisions themselves.

A number of ethical issues have been raised in international debates over the morality of patents, and there have been strong calls against the patenting of medical innovations. Laws on intellectual property vary between countries, despite attempts to harmonize these laws among industrialized countries and members of the World Trade Organization (WTO). A number of developing countries are not members of the WTO, and often the major controversies over whether a country will join WTO is related to intellectual property rights (IPR). Better solutions are required.

Practical guidance for ethics committees needs to be clarified on public health interventions. One key problem is identifying who is specifically at risk, and what the particular risk is. In vector release studies, everyone in the area may be at risk. These complex questions are made more manageable through breaking down the concerns people have into manageable areas. Defining a minimum standard of protection for research participants in trial and control populations for GMO interventions is the key point. This issue is not specific to GM vectors and pathogens, but it is crucial to consider the benefit/risk equation.

Most concerns can be the subject of better information and education. Gathering satisfactory scientific data by conducting field trials, and understanding ecological issues,[49] are the main criteria required prior to release for most people. The remaining concern, and one which is also found in scientists as well as the public, is that genetic engineering is somehow unnatural. This is an issue that needs greater social discussion. However, if presented with the threat of contracting disease, most people have few concerns about using other "unnatural" remedies such as pesticides and medical drugs. Given that most mosquitoes do not transmit disease to humans, some would argue that it is not unnatural to change a mosquito that does transmit diseases into one that does not. This is debatable.

There is a need for public opinion studies in the communities before the release, during the process of community engagement, and after the study, if we wish to really understand the opinions and concerns that people have.

Consent from Trial Participants

Recognition of the ethical principle of autonomy means that all participants need to give informed consent to an intervention that has a reasonable risk of causing harm.[50] There are significant difficulties in obtaining individual informed consent in some developing countries,[51-53] but by adequate investment of time and provision of suitable materials, it should be possible to obtain informed consent from individuals at direct risk, even though the exact cultural interpretation of the informed consent process may vary between countries.[54] There are risks of direct or indirect harm to human beings from the original pathogen-transmitting vector, so when a trial has high expected benefits, we can argue that a trial needs to be done to show that there is greatly reduced risk of harm from the modified vector. Until a trial is conducted we cannot be sure that there will be no risk and that the whole enterprise has been successful.

The risks may not just be those that arise directly from the ability of the vector to carry the target pathogen. There could be a negative impact on human health by altering the behaviour of blood-feeding insects. In the case of insects that cannot be confined to a particular population, whether they fly or float to new places, notions of "human subject" and "informed consent" need to be extended. There are basic ethical issues involved in vector collection and studies in the field. Firstly, many such studies have relied on a researcher waiting for the vector to land on a human host, and then capturing it hopefully before the vector has transmitted the pathogen to the "bait". In fact, any field studies in which human beings are exposed to the pathogens raise the question as to why some other intervention is not used in that area.

The approach developed for population genetics studies may be useful where the community and local authorities are involved in the decision-making process. Informed consent requires information to be provided, so disseminating information about the plans and progress of the project, and obtaining the consent of any person potentially affected by the release of transgenic insects, is important for the ethical conduct of research trials, whether or not national guidelines require this, or even exist. Other lessons show us that people who lack the means to express their preferences may have been abused by the lack of individual or community consent for research in anthropology[55,56] and epidemiology.[57-60]

If a study involves humans, oversight by an ethics committee or institutional review board (IRB) is necessary. In an increasing number of countries, such committees are established by law and are charged with certain legal responsibilities, typically about the conduct of research or clinical practice at local or national level. An IRB is a group of persons from a range of disciplines who meet to discuss the ethical issues of particular submitted procedures and review the benefits, risks and scientific merit of the application. The IRB usually requires that each human subject in a medical trial gives informed consent to be involved in the project. Model ethical guidelines on the establishment and procedures for an IRB have been produced by an international consultative committee for TDR. These guidelines however are not sufficient for the broad question of how to obtain informed consent for a public health intervention involving thousands of persons where the benefits are not demonstrated.

Ethics or bioethics committees include groups of people set up to adjudicate about bioethical matters. An IRB is in a sense an institutional ethics committee, but a typical IRB works through a large number of applications and often excludes the broader social discussion and representation that is seen in a regional or national bioethics committee. There are also national variations in the laws to define membership and scope of work, and terms used. The project to introduce transgenic insects will need an ethics committee with a broad overview, and specific regional ethics committees to consider the local issues.

To consider the issue at a local level, as required for obtaining appropriate informed consent, it is essential that a local ethics committee (and/or IRB if associated with an institution) open to the communities involved is established. There are cultural differences in the way informed consent should be taken.[52,61] The accepted norm in international ethical guidelines is seen for example in the modified Helsinki Declaration[62] and the draft Council for International Organizations of Medical Sciences guidelines.[63] In cases involving bilateral research collaboration, the most stringent ethical standards of the two countries should be applied. This creates problems for non-literate populations, and for populations whose common sense social assumptions are different. It is desirable that internationally agreed standards are applied, and that there are few points of difference between these standards even for simple clinical trials of drugs. The ultimate decision procedure should be decided by the local ethics committee, but international consistency and guidance will be essential.

Although the control population for the study may continue to face the same high risk of contracting the disease, recent trends in research ethics debate whether we can leave control groups without any treatment. Therefore, ethically there may need to be some other vector reduction measures given if making any interventional study in an area. While those designing ethical guidelines on placebo-controlled trials (e.g., Helsinki Declaration) were thinking of placebo controls on clinical trials of potential medical drugs, we can ask the ethical question whether researchers have an obligation to the local population to use the best available means of disease control whenever they enter an area for a study. This practically means that, as well as studying the new method, a researcher may ethically be compelled to also provide the best available proven alternative to the study population. There may be times when the provision of the proven alternative to the area of study alters the dynamics of the disease so that the results of the vector field trial differ from what the results would have been had no established alternative been provided.

Before and during the intervention, there may be privacy concerns when questionnaires are administered and personal data are stored. For public health purposes, it is essential that all information about individuals involved is linked to other data, but to ensure privacy, the data should only be identifiable to a specific person by a coding frame that is not in a computer linked to a network.

Children are therefore one of the targets of public health interventions, with presumed consent from the therapeutic imperative that they want to be involved in programmes that will avoid disease. Some compulsory vaccination programmes have faced criticism that consent is not obtained even from the surrogate decision-maker, the child's parents. In each family there may be several adults, and more children, which raises questions of whether consent is required from every individual. The local cultural norms need also to be considered. However, an appropriate mechanism may be one in which the views of everyone of reproductive age (let us call this the level of adult maturity) are gathered, and consent sought from these persons both as individuals and as a family. The agreement and understanding of children in the community should be sought through suitable materials. However, children should not be exposed to direct risk from therapeutic trials unless there is no alternative. In the case of a child living in a community that was involved in a GM vector trial, no direct risks to the human population would be expected so the consent issue is not a major hurdle. On a more positive note, children in fact could be a very powerful means to involve the community in a process of community engagement through schools. Since children are at higher risk from many of the diseases in question, they stand to benefit more, and most parents may want to be involved in the trial because of the potential benefit to their children rather than themselves.

If the trial covers an area with a local population of 100,000 persons or more, it is unrealistic and unlikely that informed consent can be given by all people in the area. There will always be some people who are against any proposition, no matter how much others value it, but the opponents cannot be moved from their houses for the period of the trial. So a procedure that is neither paternalistic nor paralytic needs to be developed. After the process of consultation and dialogue to seek informed consent, there could still be a procedure to supply additional information to all persons in the area especially to the minority who disagree. In developing countries, many may not realistically be in either a position to achieve social consensus or for persons to actually leave the area. This is not a novel issue, but common to many policy questions, and an appropriate solution needs to be developed in dialogue with each community. In the history of public health the persons who disagree usually cannot leave. Other options may be to provide additional insecticide resources to households that object to the study and are afraid of the presence of GM insects. The mechanisms for social consensus in biotechnology are not well understood in the affluent countries that have been debating GMOs, and even less is known in developing countries. Public opinion studies suggest that people may respond differently to theoretical and real situations.

Recognizing the autonomy of people as a group demands that we apply the consent model to more than isolated individuals. The introduction of GM vectors and pathogens requires community consent, so a process for seeking group consent needs to be developed for each community.[56] The question of whether every citizen has to consent to public health interventions is not a new one,[64] but with the current social transition from a paternalistic society to informed consent and informed choice, this key concern is appearing in all societies, although at different speeds.

Any initial trial may be subject to the philosophy "not in my backyard". Socially powerful persons are generally more effective at preventing trials they perceive to be risky in their area, or, conversely, at attracting social resources towards themselves and away from weaker persons in the community. Ethically it is important that risks and benefits are shared equally, and one way to ensure this would be a commitment to the local community that, if the trial is successful, the full-scale intervention would include them from the beginning. In this way, any risks borne by a local population would subsequently be rewarded by that population being the first group to benefit from the knowledge gained when the full-scale safe and effective control programme is implemented. The field trial must therefore come with a commitment to the local community that financial resources will be available and that sustainable use of the control tool will be affordable.

Regulation and Biosafety

The internationally accepted principles of risk assessment for GMOs take into account: relevant technical and scientific details of the recipient or parental organism, the donor organism(s), the vector, the insert(s) and/or characteristics of modification, the GMO, and the methods for detection and identification of the GMO including specificity, sensitivity and reliability; as well as information relating to intended use, information on location and geographical, climatic and ecological characteristics, and the foreseen health impact of the intervention.[1] The ethical principle of non-maleficence is the underlying basis for attempting to avoid harm and the regulation of human activity.

What is a particularly relevant point in the development of GM insect vectors unless it is based on sterile insect methods,[65] is that in order for a vector programme to be successful, the modification must spread throughout the wild population of a vector. This means that deliberate infection with the transgene may be the target of introducing the GMO. In order to define the parameters associated with the speed and extent of spread of the genetic modification under real conditions, extensive trials are necessary. There would need to be substantial laboratory and caged insect trial data before open release should be tested.

The International Centre for Genetic Engineering and Biotechnology (ICGEB) provides assistance in biosafety training for the development of genetic engineering in many countries.[66] Some issues also relate to the proposed Code of Conduct in Biotechnology being developed under the Commission on Genetic Resources for Food and Agriculture (CGRFA). UNDP67 and FAO generally support the development of genetic technology while considering the benefits and risks of the organisms. The capacity of countries to establish committees to adequately address ethical, social and scientific concerns needs to be strengthened.

The Scientists' Working Group on Biosafety of the Edmonds Institute[68] in Washington D.C., USA, recommended that field trials of vectors genetically engineered to reduce disease should be small scale in terms of the area of dispersal of the vector. "In the case of an anti-malaria or anti-dengue intervention, such a field trial could involve a single village or an isolated cluster of adjacent villages. No large-scale release should be attempted until the effectiveness is shown in the first trial". Thus, while there is general international consensus in the UN system that selected use of GMOs should proceed, there are groups within society that continue to be cautious. There are questions over how to balance the dispersal of GM mosquitoes that is needed for a trial versus precaution over limiting the trial geographically. There are also countries whose political regimes do not accept GMOs, and these attitudes depend on political elections, including the principle of democracy. National sovereignty should of course be respected, but GM vectors may spread beyond a national border.

The Cartagena Protocol on Biosafety to the Convention on Biological Diversity is an advance informed agreement procedure on the safe transport, handling and use of living modified organisms resulting from modern biotechnology that specifically focuses on transboundary movements of living modified organisms. The parties to this protocol agreed to ensure that "the development, handling, transport, use, transfer and release of any living modified organisms are undertaken in a manner that prevents or reduces the risks to biological diversity, taking also into account risks to human health". It was also noted that "the parties are encouraged to take into account, as appropriate, available expertise, instruments and work undertaken in international forums with competence in the area of the risks to human health".[7] In the Cartegena Protocol, "a living modified organism means any living organism that possesses a novel combination of genetic material obtained through the use of modern biotechnology. Modern biotechnology means the application of either in vitro nucleic acid techniques, including the recombinant DNA and direct injection of the nucleic acid into cells or organelles, or the fusion of cells beyond the taxonomic family, that overcome natural physiological reproductive or recombination barriers and that are not techniques used in traditional breeding and selection". This definition of a living modified organism (LMO) is now accepted in international law in general because of the Protocol. The actual term "living modified organism" is still not as widely used as "genetically modified organism", the term that has been used for two decades in academic and media debates.

One useful development of the Cartegena Protocol umbrella is the establishment of biosafety clearing houses, which are contact points in each member country. The Protocol also includes risk assessment and risk management once agreement is reached, as well as development of capacity building in biotechnology research. Many developing countries do not have the economic or scientific capacity needed to examine the products of modern biotechnology.[69] Information related to GM vectors should be linked to the same biosafety clearing houses.

Conclusion

There are a variety of ethical issues that are raised from the use of GM insects, but the most challenging may be the process of informed consent for individuals and communities. Each community or society needs to be given a chance to set consensus values on risk assessment. This two way process of community engagement is evolving and appropriate procedures for each community need to be developed. A universal minimal standard of risk assessment applicable to disease vectors needs to be defined, as diseases cross national and continental borders and so would probably GM mosquitoes.

Before field release of transgenic insects, researchers must assess all the scientific and social issues associated with GM vectors and develop safety precautions to address potential risks. The scientific and social risks should be minimized through careful design of the vector system, relevant laboratory experience, and careful choice of the site including considering appropriate social and cultural factors. Even if there are not perceived to be any realistic risks, a procedure for their evaluation should be set up so that new information can be gathered and interpreted.

This procedure may involve establishing a specialized ethical review committee under the auspices of an international body such as TDR to offer advice to researchers on the ethics of projects.

There should be prior environmental, medical and social studies for site selection, and the most appropriate site chosen on the basis of these data. Information should be exchanged as broadly as possible with community leaders, members of the local community, and the mass media. Consent should be obtained from the communities involved. Specific mechanisms to obtain individual and group consent need to be developed for public health interventions. A contingency plan for aborting a field trial needs to be developed.

Commitment to the local communities involved in field trials should be made such that they will be the first beneficiaries of more permanent use of a GM vector should results indicate that this is appropriate. Intellectual property concerns should not be barriers to implementing public health measures using GM vectors or their symbionts and/or pathogens. Prior negotiation, including possible involvement to allow access to the latest technology, is preferable to confrontation. The data should be made available to all in order to benefit from global expertise and develop international consensus. There is a need for an ongoing and active process of ethical analysis, through a variety of forums, that will provide us with the conclusions about where it is ethical to conduct these type of studies.

Ethically, we have to consider what are core ethical values for modification of nature for human needs. Modification of nature is something that all human beings have done, but balanced technology assessment of all options, past, present and future, should be examined. The ethical principle of beneficence demands action to eliminate hunger and disease. We must do this while preserving the environment for the future and respecting the cultural diversity that each community in endemic areas has.

References

1. Macer DRJ. Ethical, legal and social issues of genetically modified disease vectors in public health. UNDP/World Bank/WHO Special program for Research and Training in Tropical Diseases (TDR), Geneva 2003.
2. Azevedo E, de Moraes Marcilio Cerqueira E. Decisions in circumstances of poverty. Eubios Journal of Asian and International Bioethics 2002; 12:105-107.
3. Rawls JA, Theory of Justice. Cambridge, MA.: Belknap Press: 1971.
4. Macer DRJ. Bioethics is love of life. Christchurch, Eubios Ethics Institute. 1998.
5. Boyd A, Ratanakul P, Deepudong A. Compassion as common ground. Eubios Journal of Asian and International Bioethics 1998; 8:34-37.
6. Ho MW. Genetic engineering: dream or nightmare? London, Gateway Books: 1998.
7. CBD (Convention on Biological Diversity), Cartagena Protocol on Biosafety. 2000. Available at: http://www.biodiv.org/biosafety/protocol.asp
8. Callahan D, Jennings B. Ethics and public health: forging a strong relationship. American Journal of Public Health 2002; 92:169-176.
9. Comstock GL. Vexing nature? On the ethical case against agricultural biotechnology. Boston, Kluwer Academic: 2000.
10. Schweitzer A. The teaching of the reverence of life. London, Peter Owen: 1966.
11. Weed DL, McKeown RE. Ethics in epidemiology and public health I. Technical terms. Journal of Epidemiology and Community Health 2000; 55:857.
12. Beauchamp TL, Childress JF. Principles of biomedical ethics. 4th Ed. New York, Oxford University Press: 1994.
13. Tsai DFC. Ancient Chinese medical ethics and the four principles of biomedical ethics. Journal of Med Ethics 1999; 25:315-21.
14. Spielman A, D'Antonio M. Mosquito: the story of mankind's deadliest foe. Faber and Faber. 2001.
15. James C. Global review of commercialized transgenic crops: 2003. New York, Ithaca, International Service for the Acquisition of Agri-biotech Applications (ISAAA). 2004.
16. FDA. U.S. Food and Drug Administration, Center for Food Safety and Applied Nutrition Guidance for Industry, Voluntary Labeling Indicating Whether Foods Have or Have Not Been Developed Using Bioengineering. Draft Guidance (January 2001) 2001. <http://www.cfsan.fda.gov/~dms/biolabgu.html>

17. United Kingdom Royal Commission on Environmental Pollution, thirteenth report. The release of genetically engineered organisms to the environment. London, H.M.S.O. 1989.
18. New Zealand Royal Commission on Genetic Modification, Final report and recommendations. New Zealand. 2002. Available at: www.gmcommission.govt.nz
19. Nuffield Council on Bioethics, Genetically modified crops: the ethical and social issues. Available at: http://www.nuffieldbioethics.org/1999.
20. Holt RA, Subramanian GN, Halpein A et al. The Genome Sequence of the Malaria Mosquito Anopheles gambiae. Science 2002; 298:129-49.
21. Morel CM, Touré YT, Dobrokhotov B et al. The Mosquito Genome—a Breakthrough for Public Health Science 2002; 298:79.
22. TDR. Scientific Working Group on Insect Disease Vectors and Human Health. Geneva, WHO/ HQ, 12-16 August 2002. Geneva, TDR, document TDR/SWG/VEC/03.1, 2002.
23. Robinson AS, Franz G, Atkinson PW. Insect transgenesis and its potential role in agriculture and human health. Insect Biochem Mol Biol 2004; 34:113-120.
24. TDR Scientific Working Group on Strategic Social, Economic and Behavioural Research, 31 May-2 June 2000. Geneva, TDR, document TDR/STR/SEB/SWG/00.1, 2000.
25. Curtis CF. The case for de-emphasizing genomics in malaria control. Science 2000; 290:1508.
26. Hoffman SL. Research (genomics) is crucial to attacking malaria. Science 2000; 290:1509.
27. James AA, Morel CM, Hoffman CL et al. Present and future control of malaria. Science 2001; 291:435-6.
28. O'Brochta DA, Atkinson PW. Building the better bug. Scientific American 1998; 279:90-95.
29. Beaty BJ. Genetic manipulation of vectors: A potential novel approach for control of vector-borne diseases. Proc Nat Acad Sci (USA) 2000; 97:10295-7.
30. Macer DRJ. Bioethics for the people by the people. Christchurch, Eubios Ethics Institute. 1994.
31. Macer DRJ, MC Ng. Changing attitudes to biotechnology in Japan. Nat Biotechnol 2000; 18:945-7.
32. Singer P. Animal liberation. London, Jonathan Cape: 1976.
33. Macer D. Uncertainties about 'painless' animals, Bioethics 1989; 3:226-235.
34. Munro L. The future animal: Environmental and animal welfare perspectives on the genetic engineering of animals. Cambridge Quarterly of Healthcare Ethics 2001; 10:314-24.
35. Reiss MJ, Straughan R. Improving nature? The science and ethics of genetic engineering. Cambridge, Cambridge University Press: 1996.
36. Bruce D, Bruce A. Engineering Genesis. The Ethics of Genetic Engineering. London: Society, Religion and Technology Project, Earthscan: 1998.
37. International HapMap Consortium. Integrating ethics and science in the International HapMap Project. Nature Reviews Genetics 2004; 5:467-475.
38. Gupta A, Guha K. Tradition and conservation in Northeastern India. Eubios Journal of Asian and International Bioethics 2002; 12:15-19.
39. Hoy MA. Impact of risk analyses on pest-management programs employing transgenic arthropods. Parasitology Today 1995; 11:229-232.
40. Food and Agricultural Organization, Genetically modified organisms. Consumers, food safety and the environment. FAO Ethics Series 2, Rome, FAO. 2001.
41. Aultman KS, Walker ED, Gifford F et al. Research ethics. Managing risks of arthropod vector research. Science 2000; 288:2321-2322.
42. USDA (United States Department of Agriculture). 2002, available at: http://www.aphis.usda.gov/ biotech/arthropod/
43. Peloquin JJ, Thibault S, Staten R et al. Germ-line transformation of pink bollworm (Lepidoptera: gelechiidae) mediated by the piggyBac transposable element. Insect Molecular Biology 2000; 9:323-333.
44. Pew Initiative on Food and Biotechnology, 2004. Bugs in the System. (http://pewagbiotech.org/ research/bugs/)
45. American Committee of Medical Entomology. Arthropod containment guidelines. 2002, available at: http://www.astmh.org/subgroup/acme.html
46. Rolston H III. Conserving natural value. New York, Columbia University Press: 1994.
47. Inaba M, Macer DRJ. Attitudes to biotechnology in Japan in 2003. Eubios Journal of Asian and International Bioethics 2003; 13:78-89.
48. Durant J. Biotechnology in public: a review of recent research. London, Science Museum for The European Federation of Biotechnology: 1995.
49. Scott TW, Takken W, Knols BG et al. The ecology of genetically modified mosquitoes. Science 2002; 298:117-9.
50. Annas GK. The rights of patients. Carbondale, Southern Illinois University Press: 1989.

51. Ekunwe EO, Kessel R. Informed consent in the developing world. Hastings Center Report 1984; 14:22-24.
52. Alvarez-Castillo FA. Limiting factors impacting on voluntary first person informed consent in the Philippines. Developing World Bioethics 2002; 2:21-7.
53. Angell M. Investigators' responsibilities for human subjects in developing countries. N Engl J Med 2000; 337:847-9.
54. Nuffield Council on Bioethics, The ethics of clinical research in developing countries. 1999. Available at: http://www.nuffieldbioethics.org/
55. Fine GA. Ten lies of ethnography: Moral dilemmas of field research. J Contemp Ethnog 1993; 22:267-94.
56. Kleinman A. Ethics and experience: an anthropological approach to health equity. 1999. Available at: http://www.hsph.harvard.edu/Organizations/healthnet/HUpapers/foundations/kleinman.html
57. Capron AM. Protection of research subjects: Do special rules apply in epidemiology? Law, Medicine and Health Care 1991; 19:184-190.
58. Chee HL, El-Hamamsy L, Fleming J et al. Bioethics and human population genetics research. In Proceedings of the UNESCO International Bioethics Committee Third Session, Volume I. Paris: UNESCO 1996; 39-63.
59. Dickens BM. Issues in preparing ethical guidelines for epidemiological studies. Law, Medicine and Health Care 1991; 19:175-183.
60. Gostin L. Ethical principles for the conduct of human subject research: Population-based research and ethics. Law, Medicine and Health Care 1991; 19:191-201.
61. Levine RJ. Informed consent: Some challenges to the universal validity of the Western model. Law, Medicine and Health Care 2001; 19:207-213.
62. World Medical Association, Helsinki Declaration. 2000.
63. Council for International Organizations of Medical Sciences (CIOMS), International guidelines for ethical review of epidemiological studies. Law, Medicine and Health Care 2001; 19:247-258.
64. Kass NE. An ethics framework for public health. American Journal of Public Health, 2001; 91:1776-82.
65. Alphey L, Beard C, Billingsley B et al. Malaria control with genetically manipulated insect vectors. Science 2002; 298:119-21.
66. ICGEB (International Centre for Genetic Engineering and Biotechnology). 2002, http://www.icgeb.trieste.it/. See also UNIDO/UNEP/WHO/FAO Working Group on Biosafety, 1991; Voluntary code of conduct for the release of organisms into the environment, ICGEB Biosafety WebPages. Available at: http://www.icgeb.trieste.it/~bsafesrv/bsfcode.htm
67. UNDP, Human Development Report 2001. Making New Technologies Work for Human Development. New York, UNDP: 2001. UNEP web page, www.unep.org, 2002.
68. Edmonds Institute, The Scientists' Working Group on Biosafety, Manual for assessing ecological and human health effects of genetically engineered organisms. Washington, Edmonds Institute. 1998. See also Union of Concerned Scientists, Statements, 2001. Available at: http://www.ucsusa.org/index.html
69. Chinsembu K, Kambikambi T. Farmers' perceptions and expectations of genetic engineering in Zambia. Biotechnology and Development Monitor 2001; 47:13-14.

CHAPTER 14

Transgenic Mosquitoes for Malaria Control:
Time to Spread Out of the Scientific Arena

Christophe Boëte*

Abstract

The release of mosquitoes that are genetically modified to destroy the malaria parasite *Plasmodium falciparum* is being considered as a possible method for malaria control. If many scientific questions concerning the possible success of such a high-tech method have not received an answer yet, it seems also crucial to question the validity of such a method and also the links between technology and science in our societies.

Introduction

The use of genetically modified mosquitoes appears to be, if not an imminent, a soon to be available tool for malaria control. But will GM mosquitoes ever be real 'weapons of malaria destruction' or could their future only be confined to be lab tools of mass distraction?

GMO are part of the recent modern molecular approach to malaria control. Indeed it was decided in the early nineties, on the initiative of a small group of molecular biologists, to develop GM mosquitoes for malaria control with a 20-year work plan. This plan focused on 3 major milestones, two of them being technological and one concerning population biology issues: (1) the stable transformation of anopheline mosquitoes by 2000, (2) the engineering of a mosquito unable to carry malaria by 2005 and (3) to carry out controlled experiments to understand how to drive this genotype into wild populations by 2010. This initiative can been seen as pioneering in the use of molecular tools for malaria control, however it may also, at least partly, explain why the technological aspects of this method have received much more consideration than the ecological and epidemiological ones.[1-3] Anyway, in such a scheme, both ecologists and molecular biologists claim funds for more research, the first aiming to determine if GM mosquitoes can be successfully deployed for malaria control, the latter ones because the creation of such a mosquito requires more costly high-tech research.

One may wonder why such a technological and quite futurist approach is perceived as a viable solution for malaria control in terms of vector control. Indeed, whereas integration is nowadays seen as key for malaria control and strongly supported,[4] this method cannot be easily integrated with other vector control programs that should continue where several species transmit malaria and/or other vector-borne diseases than malaria. However, the sequencing of one of the 4 human plasmodial species *Plasmodium falciparum* and one of the numerous vectors,

*Christophe Boëte—Institut de Recherche pour le Développement, Laboratoire Génétique et Evolution des Maladies Infectieuses, 911 avenue Agropolis, B.P. 64501 34394 Montpellier Cedex 05, France; Laboratory of Entomology, Wageningen University and Research Centre, P.O. Box 8031, Binnenhaven 7, 6700 EH Wageningen, The Netherlands; Joint Malaria Programme, Kilimanjaro Christian Medical Centre, P.O. Box 2228 Moshi, Tanzania. Email: cboete@gmail.com

Genetically Modified Mosquitoes for Malaria Control, edited by Christophe Boëte.
©2006 Landes Bioscience.

Anopheles gambiae, and the resulting great academic excitement and self-congratulation have been followed by a wave of optimism concerning the use of GM mosquitoes. Optimistic models have even predicted that this solution could prove that 'once ethical and economical implications have been settled to rid the world of malaria in a short period of time'.[5] Such a declaration is probably as close to reality as the one concerning the end of hunger and sustainable development in the world with the use of GM crops. In fact, apart from purely scientific issues, the question of the use of GMO for malaria control should be placed in the context of the importance of science and technology in our societies.

Questioning the Notion of Progress

Indeed, both science and technology have been considered the source of progress in the last decades, the idea of progress being only technical and having lost its moral part.[6] If progress was considered both moral and spiritual, as well as cognitive and technical by Montaigne[7,8] in the seventeenth century, then this conception has been completely changed. Philosophers like Bacon,[9] Condorcet[10] and Saint-Simon[11] introduced a notion of progress associated with the control of Nature. With them, the world was to be transformed by men according to their will thanks to the technique, which is based on science. This conception of knowledge differs radically from the one of Plato and renders science utilitarian. Thus, in many cases progress has been associated with the amelioration of the conditions of life for humans or at least part of humanity. Science and technology are also confused with progress with the scientist as its architect. With such an attitude towards science, progress and technology, it is not surprising to notice that modern biology has been led by molecular biology. However, after a time of apparent unlimited progress, this concept of continuous progress and its potential applications was questioned and appears less obvious in the recent decades. Indeed, contemporaneous with the perverse effects of progress, leading to the questioning of its limits,[12,13] ethical preoccupations resurged and the word 'bioethics' was created.[14]

High-Tech, Malaria Control: Open Your Mind, Open the Debate

A danger inherent in following technological progress without questioning is that science can easily favour a reductionnist and mechanistic approach such as molecular biology[15] and with such a trend, high-tech solutions for malaria control are nowadays more in vogue than low-tech ones. If, with no doubt, molecular biology has led to important discoveries and shown its interest and efficacy to answer questions, it does not provide any information about the other processes of high importance in malaria control such as ecology or epidemiology and its technical explanations make difficult its diffusion to a non-initiated public. Also, quite worrying in the case of GM mosquitoes (just as in the case of GM plants in agriculture), the ecological questions or fears often appear to only be, at least according to molecular biologists, a formality that should and could be solved with adequately engineered high-tech solutions. Is this optimism, utopianism, bad faith or just lack of knowledge? Logically, ecologists have claimed the need for more research in ecology and population biology. Such a motivation is laudable if its aim lays in providing clear evidence of the validity or not of such a method and not with the motivation of getting a piece of the cake (i.e., funds invested in research concerning GM mosquitoes technology) and associated high-profile papers. However molecular biologists have already begun to do part of the 'required' ecological studies[16,17] and their conclusions, even if they are not very positive about GM mosquitoes, often claim that technological solutions can be used to solve the problem. In catching the train of GM mosquitoes, population biologists and ecologists, should try to determine the 'incompressible' factors, (especially the demographic and environmental influences on resistance, evolutionary and epidemiological outcomes) that may render any transgenic approach useless after years of investment.

Malaria Control, Science and the Social Issues

When developing an area of research and its possible impact, the social issues and consequences of its use need to be fully addressed.

Malaria: Also a Social and Political Problem

Malaria was eradicated decades ago from Europe and Northern America without any genomics information but mainly by economic and social changes driven by political will. It is difficult to disagree with Michael Ashburner[18] when he declares that it is unlikely that any changes in the malaria burden will be achieved without political will. Scientists should keep in mind that any technical progress in GM mosquitoes will lead to nothing efficient if political will and a large-scale well-organised, financed and sustained campaign does not follow.

A Trade-Off High-Tech/Low-Tech

Moreover, if genomics is sometimes seen as crucial in the fight against malaria,[19] a trade-off may occur in resources for research and implementation of control methods.[20,21] More worrying, as suggested by Rogers and Randolph about tsetse control[22] high-tech programs are likely to depend on external expertise and technology requiring huge investment upfront. Thus any failure in this chosen approach will only bring massive debt and divert funds from traditional control activities.

Obviously one may argue that high-tech and low-tech malaria control methods do not request their funding from the same sources[23] and that low-tech methods have also to face implementation. Indeed impregnated bed-nets are used by less than 2% of the population at-risk in Africa, despite the Abuja declaration stating that this coverage should be 60% in 2005. One may then wonder why, if implementation of 'efficient tools' does not work, should more 'technological' ones stand more chance?

Legal and Ethical Issues Concerning the Use of GM Mosquitoes for Malaria Control

If legal, social and ethical issues have begun to be debated[24] they should maybe be raised well before technological ones. For the moment very little attention has been turned to the possible adverse effects of the use of GM mosquito and the 'precautionary principle'. The civil society has not yet been consulted in the recent closed meetings and workshops (London 2001, Atlanta 2001, Wageningen 2002 and Nairobi 2004) on the topic. It appears highly important and needed that NGOs and community groups are involved and consulted on these topics. The development of infrastructures and technology related to GM mosquitoes in Africa is said to be of high importance,[25] however it has to be accompanied by the development of organisations able to be critically engaged in science and technology policy both in Western countries and in malaria-endemic areas. This is an essential part of democratizing science and technology.

Democratizing Science

As discussed in a document from ITDG (Intermediate Technology Development Group),[26]

> *"The development process for most new technologies still uses a model unchanged since the nineteenth century—first, optimise the technology, then check user acceptance, and finally examine any regulations governing its use. Given the investments made in the earlier stages, it becomes difficult to redesign a technology even when potentially harmful social effects have been subsequently identified. Hence, when faced with opposition to a new technology, policy-makers are forced into defending the technology, a technocratic managerial response in which potential social and environmental impacts, identified outside the narrow design process, are regarded as problems of user acceptance".*

One might indeed regret the absence of engagement of non-specialists both in discussing the research funding priorities but also in the choice of any technological process. If it might be argued that nonspecialists could not be included in those decisions because of a lack of knowledge, examples exist where non-specialists have been participating in decision-making and also shaping the orientation of science and technology.[26] It happens with the Quality Research in Dementia network in England where donors of the Alzheimer's Society (a major charity) discuss grant proposals from the scientists. Concerning GMO, the Movimento dos Trabalhadores Rurais Sem Terra (MST - Landless Workers Movement) is leading a programme of rural land reform, and by promoting low-input agriculture and low cost agro-ecological technologies it has led to major set-backs for Monsanto as this latter one attempts to introduce GM crops into Brazil. In both cases, a collective rather than a hierarchical control occurs and scientific and technological aspects of research are under the control of citizens. To adopt the same attitude with research on transgenic mosquitoes for malaria control would oblige scientists to give a jargon-free description of their proposed research and permit the citizen to evaluate projects and their potential repercussions, which should be welcome by scientists as they'll have the feed-back from the citizens that may profit from their so-called applications.

Conclusion

There has been too much false hope for malaria endemic areas (the eradication attempt, the sporozoite-irradiated vaccine in the 1960s) that has led to stinging failures. The actual most important tools to reduce the malaria burden are not really cutting edge high-tech methods (they also do not have the associated pizzaz): the proper use and free distribution of impregnated bednets,[27,28] the use of derivatives of artemisine and house improvement.[29] It also appears fundamental to improve the development of logistically and financially facilitated access to health care infrastructures where patients can get healthcare of proximity and quality,[30] the colossal needs of the health sector in many African countries are indeed sadly associated with the deliquescence of states and the associated Structural Adjustment Programs and liberal politics.[31,32]

Finally, the question that arises regards the effect of malaria research on the malaria situation and consequently does malaria research have to be applied to justify itself? With no doubt any research on malaria will lead to biological discoveries that may be associated with unforeseen results. Using malaria parasites and mosquitoes as research systems should certainly not be considered as applied research with an associated need of (sometimes) far-fetched excuses or false pretences for short-term applications as GM mosquitoes are probably raising more questions in fundamental research than in malaria control.[33]

One might also take care that science does not become a technoscience,[34] being restrained to a means of the technique. Moreover, the necessity of applied outcomes makes research at risk of losing its freedom and a science able to lead to technological innovations may supplant a science without function apart from satisfying the intellectual curiosity of mankind. Thus it is useful to recall Erwin Chargaff who thought that the idea that science can improve the world is hubris. What is needed from members of the scientific community is honesty about their motivations and the plausible benefits of their research to humanity. This, along with a reserve towards their recent exciting discoveries, should be the most laudable attitude of scientists. Malaria research has a central position in today's research, and this gives scientists a very powerful position. They can use it to advise decision-makers and, especially if they truly consider their research as potentially applied, they cannot avoid social and societal debates and forget their responsibility as citizens.

Acknowledgements

I thank Caroline Ellson, Daniel Keates and Florence Rivenet for helpful discussion on previous versions of this manuscript.

Reference

1. Boëte C, Koella JC. Evolutionary ideas about genetically manipulated mosquitoes and malaria control. Trends Parasitol 2003; 19(1):32-38.
2. Takken W, Scott TW, eds. Ecological Aspects for Application of Genetically Modified Mosquitoes. Vol. 2. Dordrecht, The Netherlands: Kluwer Academic Publishers, 2003.
3. Boëte C. Malaria parasites in mosquitoes: Laboratory models, evolutionary temptation and the real world. Trends Parasitol 2005; 21(10):445-447.
4. Utzinger J, Tanner M, Kammen DM et al. Integrated programme is key to malaria control. Nature 2002; 419(6906):431.
5. Hahn MW, Nuzhdin SV. The fixation of malaria refractoriness in mosquitoes. Curr Biol 2004; 14:R264-R265.
6. Bellon A. Des savants parfois schizophrènes. Le Monde Diplomatique 2002; 26.
7. Montaigne M. Essais, 1580.
8. Larmat J. l'idée de progrès dans les Essais de Montaigne. Paris: Vrin, 1982.
9. Bacon F. Novum organum 1620.
10. Condorcet. Esquisse d'un tableau historique du progrès de l'esprit humain. 1793.
11. Saint Simon. Lettres d'un habitant de Genève à ses contemporains. Œuvres de Saint-Simon 1803.
12. Jonas H. Le principe responsabilité. Un éthique pour la civilisation technologique. Paris: 1979.
13. Taguieff PA. Le sens du progrès une approche historique et philosophique. Paris: 2004.
14. van Rensselaer P. Bioethics: Bridge to the Future. NJ: Prentice Hall, 1971.
15. Gouyon PH. Pas de progrès sans raison ni précaution. POUR La revue du groupe de recherche pour l'éducation et la prospective. Dossier Sciences and Agriculture, Accord et Désaccords 2003; 146-151.
16. Catteruccia F, Godfray HCJ, Crisanti A. Impact of genetic manipulation on the fitness of anopheles stephensi mosquitoes. Science 2003; 299(5610):1225-1227.
17. Moreira LA, Wang J, Collins FH et al. Fitness of anopheline mosquitoes expressing transgenes that inhibit Plasmodium development. Genetics 2004; (166):1337-1341.
18. Ashburner M. A hat trick - Plasmodium, Anopheles and Homo. Genome Biol 2002; 4(1):103.
19. Hoffman SL. Infectious disease: Research (Genomics) is crucial to attacking malaria. Science 2000; 290(5496):1509.
20. Curtis CF. Infectious disease. The case for deemphasizing genomics in malaria control. Science 2000; 290(5496):1508.
21. Smith SM. High-and low-tech malaria control. Science 2004; 304:1744.
22. Rogers DJ, Randolph SE. A response to the aim of eradicating tsetse from Africa. Trends Parasitol 2003; 18(12):534-536.
23. Hemingway J, Craig A. Reply to Smith SM. Science 2004; 304:1744.
24. Touré YT, Oduola AMJ, Sommerfeld J et al. Biosafety and risk assessment in the use of genetically modified mosquitoes for disease control. In: Takken W, Scott TW, eds. Ecological Aspects for Application of Genetically Modified Mosquitoes. Dordrecht: Kluwer Academic Publishers, 2003:9-12.
25. Mshinda H, Killeen GF, Mukabana WR et al. Development of genetically modified mosquitoes in Africa. Lancet Infect Dis 2004; 4:264-265.
26. Wakeford T. Democratising technology Reclaiming science for sustainable development: ITDG. 2004.
27. Curtis C, Maxwell C, Lemnge M et al. Scaling-up coverage with insecticide-treated nets against malaria in Africa: Who should pay? Lancet Infect Dis 2003; 3(5):304-307.
28. Hinh TD. Use of insecticide-impregnated bed nets for malaria vector control in Vietnam. 2002.
29. Lindsay SW, Emerson PM, Charlwood JD. Reducing malaria by mosquito-proofing houses. Trends Parasitol 2002; 18(11):510-514.
30. Olivier de Sardan JP. Une médecine de proximité... et de qualité pour l'Afrique. Le Monde Diplomatique 2004; 18-19.
31. Olivier de Sardan JP. Dramatique déliquescence des Etats en Afrique. Le Monde Diplomatique 2000; 12-13.
32. Castro J, Millet D. Malaria and structural adjustment: proof by contradiction. In: Boëte C, ed. Malaria Control with Genetically Modified Mosquitoes. Georgetown: Landes Bioscience, 2006; 2:16-23.
33. Chevillon C, Paul REL, de Meeus T et al. Thinking transgenic vectors in a population context: Some expectations and many open-questions. In: Boëte C, ed. Genetically Modified Mosquitoes for Malaria Control. Georgetown: Landes Bioscience, 2006; 10:117-136.
34. Ellul J. La technique ou l'enjeu du siècle. Paris: Armand Colin, 1954.

Index

P

Parasite development 9, 24, 28, 30, 31, 37,
105, 107, 111, 125, 127, 138
Partnership 2, 147, 149, 150
Physiological cost 125
Plasmodial species 83, 84, 137, 138, 166
Plasmodium falciparum 1, 4, 6-9, 37, 40, 55,
79, 80, 83-86, 89, 90, 92-94, 96, 98, 99,
108, 128-132, 166
Polymorphism 6, 7, 36-42, 47, 48, 51, 52,
124, 142
Population 2, 5, 6, 8-10, 16, 19-22, 24, 26,
31, 32, 39-41, 46-55, 60-62, 64, 66-71,
73, 74, 76, 79-81, 83, 84, 86, 89-92, 94,
97, 98, 103, 104, 108-113, 117-122,
124, 126-130, 133, 134, 137-140, 142,
143, 149, 150, 155, 156, 158-161,
166-168
Population genetics 40, 46, 47, 52-54, 61,
117, 118, 129, 159
Poverty 4, 17, 21, 22, 152
Precautionary principle 153, 168
Prevalence 2, 21, 54, 55, 80, 81, 83, 90, 91,
94, 96, 103-106, 111, 112, 129, 131
Prevention 4, 10, 17, 98, 142, 146, 152
Progress 1, 5, 7-10, 25, 55, 79, 146, 149, 154,
159, 167, 168
Public concerns 147-149
Public health 146-150
Public opinion 155, 158, 159, 161

R

Refractoriness 46, 48, 49, 54, 62, 66-70, 73,
76, 79-81, 83, 84, 86, 89, 103, 104, 109,
110, 112, 113, 140
Reglobalisation 16, 17
Replacement 24, 39-41, 60, 129, 142
Resistance 1, 2, 4-6, 10, 24-26, 32, 36, 37,
40-42, 48, 62, 79-81, 83, 84, 86, 90,
103, 104, 109, 117-119, 121-131, 133,
134, 137, 146, 153, 167
Risk 2-8, 10, 54, 89-91, 95, 96, 98, 107, 108,
118-121, 123, 124, 130, 133, 147-150,
152, 153, 156-162, 168, 169
Risk assessment 161, 162
RNA interference (RNAi) 24, 28-31

S

Safety assessment 147, 148
Science 7, 8, 134, 142, 148, 153, 158, 160,
166-169
Selectable marker 24, 26, 27
Selection 5, 6, 8, 39-42, 48, 50, 52, 53, 67,
68, 70, 84, 86, 104, 107, 109, 110, 112,
113, 117, 119, 121, 125, 129, 130, 133,
142, 143, 147, 162, 163
Severe malaria 3, 4, 6, 9, 10, 90, 92-95, 98,
129
Simulation 36, 60
Social issues 148-150, 162, 168
Society 51, 113, 152-155, 157, 158, 161,
162, 166-169
Spread 1, 30, 42, 46-50, 52-55, 66, 68, 69,
73, 79-81, 83, 86, 98, 103, 104, 138,
146, 147, 155, 157, 161, 162, 166
Sterile insect technique (SIT) 31, 32, 60-62,
64, 66
Sterile males 31, 32, 60-62, 64, 66, 118
Structural adjustment 16-22, 169
Survival 36, 41, 62, 84, 104-107, 111-113,
137, 139, 143
Susceptibility 5-7, 10, 40, 69, 73, 105, 113,
117, 123, 126, 129, 130, 138

T

Technology 10, 19, 24, 26, 28, 31, 32, 36, 62,
89, 137, 138, 142, 143, 147, 152, 153,
161, 163, 166-169
Technoscience 169
TEP 30, 37, 41
Transgene 30, 31, 46-49, 54, 55, 60, 69, 70,
73, 79, 103, 118-123, 125, 126, 128,
133, 134, 157, 161
Transgenic 10, 24-26, 29-32, 47, 54, 55,
60-64, 67, 69-71, 75, 76, 79, 80, 82, 83,
86, 89, 98, 99, 103, 117-125, 129-134,
140, 146, 147, 159, 162, 166, 167, 169
Transmission 1-3, 5, 8-10, 17, 31, 46-48, 54,
55, 60, 61, 68, 79-83, 86, 89-99, 103,
105-112, 117, 126-128, 130-134,
137-142, 146, 147, 149, 150
Transposable element 24-28, 62, 69, 80, 83,
155
Transposon 25-28, 60, 62-65, 69, 70, 73, 74,
76, 121, 123, 155

V

W